Stephan Camerer

Interfacing ultracold atoms and mechanical oscillators

Stephan Camerer

Interfacing ultracold atoms and mechanical oscillators

Südwestdeutscher Verlag für Hochschulschriften

Impressum/Imprint (nur für Deutschland/only for Germany)
Bibliografische Information der Deutschen Nationalbibliothek: Die Deutsche Nationalbibliothek verzeichnet diese Publikation in der Deutschen Nationalbibliografie; detaillierte bibliografische Daten sind im Internet über http://dnb.d-nb.de abrufbar.
Alle in diesem Buch genannten Marken und Produktnamen unterliegen warenzeichen-, marken- oder patentrechtlichem Schutz bzw. sind Warenzeichen oder eingetragene Warenzeichen der jeweiligen Inhaber. Die Wiedergabe von Marken, Produktnamen, Gebrauchsnamen, Handelsnamen, Warenbezeichnungen u.s.w. in diesem Werk berechtigt auch ohne besondere Kennzeichnung nicht zu der Annahme, dass solche Namen im Sinne der Warenzeichen- und Markenschutzgesetzgebung als frei zu betrachten wären und daher von jedermann benutzt werden dürften.

Verlag: Südwestdeutscher Verlag für Hochschulschriften GmbH & Co. KG
Dudweiler Landstr. 99, 66123 Saarbrücken, Deutschland
Telefon +49 681 37 20 271-1, Telefax +49 681 37 20 271-0
Email: info@svh-verlag.de

Zugl.: München, LMU, Diss., 2011

Herstellung in Deutschland:
Schaltungsdienst Lange o.H.G., Berlin
Books on Demand GmbH, Norderstedt
Reha GmbH, Saarbrücken
Amazon Distribution GmbH, Leipzig
ISBN: 978-3-8381-0821-6

Imprint (only for USA, GB)
Bibliographic information published by the Deutsche Nationalbibliothek: The Deutsche Nationalbibliothek lists this publication in the Deutsche Nationalbibliografie; detailed bibliographic data are available in the Internet at http://dnb.d-nb.de.
Any brand names and product names mentioned in this book are subject to trademark, brand or patent protection and are trademarks or registered trademarks of their respective holders. The use of brand names, product names, common names, trade names, product descriptions etc. even without a particular marking in this works is in no way to be construed to mean that such names may be regarded as unrestricted in respect of trademark and brand protection legislation and could thus be used by anyone.

Publisher: Südwestdeutscher Verlag für Hochschulschriften GmbH & Co. KG
Dudweiler Landstr. 99, 66123 Saarbrücken, Germany
Phone +49 681 37 20 271-1, Fax +49 681 37 20 271-0
Email: info@svh-verlag.de

Printed in the U.S.A.
Printed in the U.K. by (see last page)
ISBN: 978-3-8381-0821-6

Copyright © 2011 by the author and Südwestdeutscher Verlag für Hochschulschriften GmbH & Co. KG and licensors
All rights reserved. Saarbrücken 2011

Zusammenfassung

In dieser Dissertation stelle ich experimentelle Untersuchungen zur Kopplung zwischen mechanischen Oszillatoren und ultrakalten Atomen vor. Insgesamt werden drei Kopplungsmechanismen untersucht.

In einem ersten Experiment wird unter Ausnutzung des Oberflächenpotentials eines mechanischen Oszillators dessen Bewegung an die Schwerpunktsbewegung eines Bose-Einstein Kondensates gekoppelt. Die Tiefe des magnetischen Fallenpotentials wird in der Nähe des Oszillators von dessen Oberflächenpotential reduziert. Auslenkung des Oszillators führt zu einer Modulation von Frequenz und Minimumsposition der magnetischen Falle. Der Atomzahlverlust durch die Kopplung wird in Absorptionsabbildung bestimmt, und ermöglicht die Amplitude des Oszillators mit den Atomen auszulesen.

In einem zweiten Experiment untersuchen wir die Kopplung zwischen einem mechanischen Membran-Oszillator und optisch gefangenen thermischen Atomen. Die Membran ist der Endspiegel eines optischen Gitters und die Oszillationsbewegung der Membran koppelt über dieses an die Schwerpunktsbewegung der Atome. Umgekehrt verteilen die Atome Photonen zwischen den beiden laufenden Wellen um, die das Gitter formen, wodurch die Leistung der laufenden Wellen, und letztlich der auf die Membran wirkende Strahlungsdruck moduliert wird. Wir beobachten in Absorptionsabbildung die *actio* der oszillierenden Membran auf die Atome als Temperaturerhöhung. Um die *reactio* der Atome auf die Membran nachzuweisen, wird die Dämpfungsrate der Membran mit und ohne im Gitter gefangenen Atome gemessen. In Übereinstimmung mit den theoretischen Erwartungen messen wir eine durch die Atome erhöhte Dämpfungsrate. Dieses Experiment ist der erstmalige experimentelle Nachweis der *reactio* eines atomaren Ensembles auf einen mechanischen Oszillator.

Wir untersuchen ein drittes Kopplungsschema, bei dem die Bewegung eines mechanischen Oszillators an den kollektiven Spin eines Bose-Einstein Kondensates gekoppelt wird. Die Spitze eines mechanischen Oszillators ist mit einem Magneten funktionalisiert, der dessen Bewegung in Oszillationen eines magnetischen Feldes übersetzt. Jene überführen gefangene Atome durch Umklappen des atomaren Spins in ungefangene Bewegungszustände. Die Kopplungsstärke ist hier nicht wie in den anderen Kopplungsschemata durch die Wurzel aus dem Massenverhältnis von Atomen und Oszillator beschränkt. Wir untersuchen dieses Kopplungsschema theoretisch, und diskutieren eine mögliche Realisierung eines Nanoresonators mit magnetischer Insel. Ich gebe einen Überblick über den Status der Fabrikation, und schlage eine vereinfachte Fabrikationsmethode vor.

Abstract

In this thesis I present experiments investigating controlled coupling between mechanical oscillators and ultracold atoms. I report on three different coupling mechanisms.

In a first experiment, the surface potential experienced by atoms close to the mechanical oscillator is employed to couple the oscillator motion to the center of mass (COM) motion of a trapped Bose-Einstein condensate (BEC). The magnetic trapping potential is modified by the surface potential arising from the oscillator surface which results in a reduced trap depth. Vibration of the oscillator leads to a modulation of the trap frequency and the minimum of the trapping potential. Observing the loss of atoms from the BEC allows us to read out the amplitude of the mechanical oscillator with the atoms.

In a second experiment, we study the coupling of a mechanical membrane oscillator and thermal atoms trapped in a 1D optical lattice. The membrane is the end mirror of the lattice, and oscillation of the membrane couples to the COM mode of the atomic ensemble. Conversely, the center of mass motion of the atomic ensemble redistributes photons between the two running waves forming the 1D optical lattice, effectively modulating their power, and hence the radiation pressure acting onto the membrane. We observe the action of the oscillating membrane onto the atoms by detecting the resulting temperature increase of the atomic ensemble in absorption imaging. To observe the backaction of the atoms onto the mechanical oscillator, the mechanical damping is measured in experiments with and without atoms in the lattice, and we measure higher damping in the presence of atoms in agreement with the theoretical predictions. These experiments are the first demonstration of backaction of an atomic system onto a mechanical oscillator.

We investigate a third coupling mechanism, where the motion of a mechanical oscillator is coupled to the collective spin of a BEC. The tip of a mechanical oscillator is functionalized with a magnet, which transduces the oscillators' motion into oscillations of the magnetic field. This drives spin-flip transitions of trapped atoms to untrapped motional states. The coupling strength is not limited by the square root of the mass ratio of atoms and oscillator as in the other coupling schemes discussed in this thesis. We investigate this coupling scheme theoretically, and discuss the realization of a nanometer-sized mechanical oscillator with a magnetic island. I report on the status of the fabrication, and propose a simplified fabrication method.

Contents

1 Introduction   **1**

2 Protagonists and Antagonists   **7**
    2.1 Mechanical oscillators 7
        2.1.1 Analytical description 7
        2.1.2 Excitation, damping and thermal motion 13
        2.1.3 Towards quantum mechanical oscillators 16
    2.2 Ultracold atoms 18
        2.2.1 Magnetic trapping on atoms chips 19
        2.2.2 Optical trapping 25

3 Mechanical coupling via the surface potential   **31**
    3.1 Coupling scheme 31
        3.1.1 Surface potentials 32
        3.1.2 Effect of the surface potential onto trapped atoms 33
        3.1.3 Mechanical modes of Bose-Einstein condensates 35
    3.2 Atom chip setup 37
        3.2.1 'Standard' setup 37
        3.2.2 Atom chip and BEC production 38
    3.3 Measurements 40
        3.3.1 Characterization of the surface potential 40
        3.3.2 Imaging of the mechanical oscillators' resonance 42
        3.3.3 Spectroscopy of the trapped BEC 45

4 Optomechanical coupling via an optical lattice   **47**
    4.1 Coupling scheme 47
        4.1.1 Effect of the mechanical oscillator onto trapped atoms 48
        4.1.2 Effect of the atoms onto the mechanical oscillator 51
        4.1.3 Backaction of the atoms onto the damping of the membrane . 53
        4.1.4 Coupled system – fully quantized theory 56
    4.2 Extension of the atom chip setup 60
        4.2.1 Michelson interferometer for membrane readout 60
        4.2.2 Controlled excitation of the membrane amplitude 66
        4.2.3 Optical lattice setup 67

4.3		Experimental results	71
	4.3.1	Characteristics of the SiN membrane	71
	4.3.2	Properties of atoms in the optical lattice	76
	4.3.3	Effect of the membrane onto trapped atoms	78
	4.3.4	Backaction of the atoms onto the membrane	83

5 Coupling to the collective spin of a BEC — 87

5.1		Coupling scheme	87
5.2		Design for achieving a strong coupling	89
	5.2.1	Large thermal amplitude x_{rms}	90
	5.2.2	Large magnetic field gradient G_m	90
	5.2.3	Distortion of the magnetic trap	92
	5.2.4	Trapping of atoms	93
5.3		Resolving thermal motion	93
5.4		Mechanical cavity QED	96
5.5		Steps towards the experimental realization	98
	5.5.1	Fabrication methods	100
	5.5.2	Process flow	102
	5.5.3	Characterization	108
	5.5.4	Fabrication: Status	110
	5.5.5	Fabrication: Perspective	113

6 Outlook — 116

Appendix: Parameters of chip fabrication — 120

Bibliography — I

1 Introduction

The concept of a harmonic oscillator is one of the building blocks of physics and appears in various contexts, ranging from atomic to solid state physics, and many other fields. The properties of harmonic oscillators are often visualized with mechanical oscillators, where a harmonically bound mass oscillates around a potential minimum. However, mechanical oscillators are not only used as an example sytem to study the properties of the harmonic oscillator model, but constitute a vivid field of research themselves. Many applications of micro- and nanosized mechanical oscillators have been investigated in recent years due to their high sensitivity with respect to forces [1, 2], temperature changes [3] or the variation of the oscillating mass [4]. Even the spin of a single electron in a solid can be detected with magnetic resonance force microscopy [5].

The mechanical modes of such oscillators are highly occupied with phonons in a room temperature environment, and the behaviour of such oscillators can be described classically. However, many experiments [6, 7, 8, 9] have recently achieved a significant reduction of the phonon number occupation of mechanical modes, and are on the way to 'cool' the motion of a single mode of a mechanical oscillator to the quantum ground state. In contrast to the more traditional approach, where all mechanical modes are simultaneously cooled by reducing the thermal bath temperature [10, 11, 12, 13, 14], optomechanical cooling applies a technique which is similar to laser cooling of atoms [15, 16]. In the optomechanical cooling experiments [6, 7, 8, 9], mechanical oscillators of micron size are integrated into optical cavities, and one exploits the resonance characteristics of optical cavities in order to couple the light field to the mechanical displacement of the oscillator via radiation pressure. This allows one to reduce the initial phonon number occupation of a single mechanical mode.

To study quantum mechanics with a massive mechanical oscillator beyond the process of ground state cooling, it would be favourable to manipulate and to read out the oscillators' quantum state experimentally. In an optomechanical system, squeezed light could be used to prepare a squeezed state of the mechanical oscillator [17]. Moreover, the degree of control which is already achieved for two-level quantum systems such as atoms [18, 19, 20, 21], ions [22] or superconducting flux qubits [23, 24] could be extended to mechanical oscillators close to the ground state by coupling the mechanical oscillator to such a two-level system. A coupled system would be an implementation of the Jaynes-Cummings model [25], where e.g. a neutral atom takes the role of the two-level system, and a single mode of the mechanical oscillator takes the role of the bosonic field mode, similar to the cavity light field in cavity quantum electrodynamics. The coupling of the mechanical oscillator to the two-level system is of resonant character such that the coupling can be controlled and switched off by tuning the frequency of either of the two coupled systems. If the coupling is sufficiently strong, a superposition state can be prepared in a two-level system and swapped to a mechanical oscillator in order to engineer a non-classical state in a macroscopic mechanical system.

The coupling of solid state quantum systems such as single electron transistors, Cooper pair boxes or superconducting flux qubits to mechanical oscillators is currently investigated in several experiments [10, 11, 12, 13]. Recently, strong coupling of a solid state quantum system to a dilatational volume mode of a mechanical oscillator in the groundstate was demonstrated on the single phonon level [14]. The operation of such quantum systems requires cryogenic temperatures to reduce the decoherence which arises from the relatively strong coupling to the environment. This results also in relatively short coherence times of solid state quantum systems. However, this is compensated in the experiments by fast coupling rates to mechanical oscillators.

In contrast, atomic quantum systems profit from an excellent isolation from the environment and stand out with coherence times of several tens of seconds [26]. Inspired by the experimental success in cooling a mode of a mechanical oscillator, a variety of theoretical proposals [27, 28, 29, 30, 31, 32, 33, 34, 35, 36, 37, 38, 39, 40] suggests to couple mechanical oscillators and atomic quantum systems, most of them claiming the feasibility of strong coupling with realistic parameters. Despite of these numerous proposals, there are only few experiments which demonstrate the coupling of a mechanical oscillator to an atomic system. In a first experiment [41], the atomic spin of atoms in a room temperature vapor cell was coupled to the motion of a mechanical oscillator via a magnetic gradient field, which arises from a magnet at the oscillators' tip. In this implementation, the control over the coupling is limited due to the thermal motion of the atoms in the vapour cell. While the atoms were used to detect the motion of the mechanical oscillator, the backaction of the atoms onto the oscillator was not observed in this experiment.

This thesis

In this thesis, I present experiments investigating the controlled coupling between mechanical oscillators and ultracold atoms. We study three different coupling mechanisms. First, I report on an experiment where the surface potential experienced by atoms close to the mechanical oscillator is employed to couple the oscillator motion to the center of mass (COM) motion of a trapped Bose-Einstein condensate (BEC). We resolve the mechanical resonance of the driven oscillator with the atoms, and use the coupling to perform a spectroscopy of the trapped BEC.

In a second experiment, the motion of a mechanical membrane oscillator and the atomic COM motion are coupled via a 1D optical lattice. In this experiment, we observe both the effect of the membrane onto the atoms as well as the backaction of the COM motion of the atomic ensemble onto the membrane oscillations. The backaction of laser cooled atoms leads to an increased mechanical damping rate of the oscillator, an effect which could be used for sympathetic cooling of the oscillator via the atoms.

In these experiments, the coupling strength between the two systems shows a scaling that is characteristic for the direct coupling of two mechanical oscillators via a distance dependent force. The coupling strength depends on the square root of the ratio of their masses. For coupled oscillators with very different masses, one expects a slower energy exchange rate than for equal masses. This limitation is circumvented in the third coupling mechanism that I present in this thesis. In this scheme, the mechanical oscillator is coupled to the collective spin of an atomic Bose-Einstein condensate rather than to its motion. We investigate this mechanism theoretically. I discuss the theoretical proposal and steps towards its experimental realization.

Coupling via the surface potential We use an atom chip to prepare and manipulate ultracold atoms in a magnetic trapping potential. Atom chips provide a robust toolbox to achieve [42, 43] and study [44, 45, 21, 46] BEC in compact setups. BECs can be trapped close to surfaces without compromising on coherence times [47]. We ramp trapped ^{87}Rb atoms to a distance $d < 1$ μm from a mechanical oscillator, which is glued onto an atom chip. The magnetic trapping potential is modified by the surface potential arising from the non-functionalized oscillator surface which results in a reduced trap depth [48]. The deformation of the trapping potential leads to a loss of atoms from the BEC and is measured with absorption imaging [49] by varying the distance d to the undriven mechanical oscillator. This measurement is performed on both sides of the oscillator and allows one to calibrate the distance d. Since the position of a magnetic trap is referenced to the wires on the atom chip, we can position a BEC reproducibly with a resolution better than 7nm at a distance of $d = 1.3$ μm. Vibration of the oscillator leads to a modulation of the trap frequency and the minimum of the trapping potential. This results in a coupling of the motion of the driven mechanical oscillator to the motion of trapped atoms. This coupling allows us to read out the amplitude of the mechanical oscillator with the atoms, and to resolve the fundamental mode resonance. It is further employed to perform a spectroscopy of the trapped BEC. We demonstrate controlled excitation of the center of mass mode, the radial breathing mode and the quadrupole mode, which occurs at this frequency only in a BEC due to the atomic mean field interaction. In this experiment, the surface potential is used to measure the action of the mechanical oscillator onto the atomic ensemble. The backaction of the atoms onto the mechanical oscillator is simply too small to be measured in this experiment.

Coupling via a 1D optical lattice In the second experiment, we observe the backaction of an atomic ensemble onto the mechanical oscillator. The coupling mechanism was investigated theoretically in a collaboration with Klemes Hammerer *et. al.* in the group of Peter Zoller, and published in [39]. In the experiment, a red detuned laser beam is reflected at a mechanical membrane oscillator in order to provide a 1D optical lattice potential for ultracold atoms. Oscillations of the membranes' fundamental mode shift the lattice potential and couple to the COM mode of the atomic ensemble. Conversely, the center of mass motion of the atomic ensemble redistributes photons between the two running waves forming the 1D optical lattice, and effectively modulates their power, and hence the radiation pressure acting onto the membrane. The mechanical oscillator used in the experiment is a low-stress SiN membrane with a mechanical quality factor of 1.5×10^6 at an eigenfrequency of 273 kHz in our room temperature setup. The calculated amplitude reflectivity $R = 0.56$ of the bare silicon nitride allows one to use the membrane itself as a lattice end mirror, and results in an asymmetric coupling between the mechanical oscillator and the atomic ensemble, which is typical for cascaded quantum systems [50, 51]. We observe the action of the oscillating membrane onto the atoms by detecting the resulting temperature increase of the atomic ensemble. The membrane excites the center of mass mode, and the broadening of the cloud due to dephasing of the COM mode in the anharmonic lattice potential is observed in time of flight measurements. To observe the action of the atoms onto the mechanical oscillator, the lattice is permanently replenished with atoms from a steady state magneto-optical trap. The membrane is coherently excited to a well-defined amplitude, and the decay of the amplitude is monitored after the membrane excitation is interrupted. The decay times observed in such ringdown measurements with and without atoms are compared, and we find shorter time constants in the presence of atoms, as expected. We find a reasonable agreement of the experimental results and the theoretical predictions. These experiments are the first demonstration of backaction of an atomic system onto a mechanical oscillator.

Coupling to the atomic spin Coupling the motion of a mechanical oscillator to the collective spin of an atomic ensemble bears the advantage that the coupling strength is not limited by the square root of the mass ratio as in the motion-to-motion coupling discussed above. In the situation that we consider, the tip of a mechanical oscillator is functionalized with a magnet, which transduces the oscillators' motion into oscillations of the magnetic gradient field. The oscillations drive spin-flip transitions in the magnetically trapped and spin polarized atomic ensemble to untrapped motional states. We investigate this coupling scheme theoretically, and give guidelines for the design of a suitable nano-sized mechanical oscillator which is functionalized with a single-domain magnet. I report on the status of the fabrication of such an oscillator, and the characterization of mechanical and magnetic properties. In addition, I propose a simplified method to functionalize nanometer-sized mechanical oscillators with single domain magnets.

Organization of the chapters

- Chapter 2 'Protagonists and Antagonists' provides the basic theory of the experimental systems interacting in this thesis. The eigenmodes of the mechanical oscillators are characterized analytically, and their excitation and dissipation mechanisms are reviewed. The trapping of neutral, ultracold atoms in magnetic traps on atoms chips or in optical potentials is described.

- Chapter 3 gives a short overview of theory, setup and experimental results of the experiment, where the surface potential is employed to couple the motion of a mechanical oscillator to magnetically trapped atoms. The mechanical motion of the driven oscillator is resolved with a BEC, and used in turn to perform a spectroscopy of the trapped BEC.

- Chapter 4 describes the theory, setup and experimental results of the experiment, where the motion of a mechanical membrane oscillator and the atomic COM motion are coupled via a 1D optical lattice. The experimental results show the backaction of ultracold atoms onto the mechanical membrane oscillator as an increased mechanical damping rate.

- Chapter 5 describes a coupling scheme theoretically, where the motion of a mechanical oscillator is coupled to the collective spin of an atomic Bose-Einstein condensate. The considerations leading to a suitable chip design of the nanofabricated chip structure, and the fabrication process developed accordingly are described in detail.

List of papers

- S. Camerer, M. Korppi, D. Hunger, A. Jöckel, T.W. Hänsch, and P. Treutlein,
  Optomechanical interface of ultracold atoms and a membrane,
  to be submitted.

- D. Hunger, S. Camerer, M. Korppi, A. Jöckel, T.W. Hänsch, and P. Treutlein,
  Coupling ultracold atoms to mechanical oscillators,
  to be published in Comptes Rendus Physique.

- K. Hammerer, K. Stannigel, C. Genes, and P. Zoller, P. Treutlein, S. Camerer, D. Hunger, and T. W. Hänsch,
  Optical Lattices with Micromechanical Mirrors,
  Phys. Rev. A **82**, 021803 (2010).

- D. Hunger, S. Camerer, T. W. Hänsch, D. König, J. P. Kotthaus, Jakob Reichel, and Philipp Treutlein,
  Resonant Coupling of a Bose-Einstein Condensate to a Micromechanical Oscillator,
  Phys. Rev. Lett. **104**, 143002 (2010).

- P. Treutlein, D. Hunger, S. Camerer, T. W. Hänsch, and J. Reichel,
  Bose-Einstein Condensate Coupled to a Nanomechanical Resonator on an Atom Chip,
  Phys. Rev. Lett. **99**, 140403 (2007).

- I. Favero, C. Metzger, S. Camerer, D. Koenig, H. Lorenz, J. P. Kotthaus, and K. Karrai,
  Optical cooling of a micromirror of wavelength size,
  Applied Physics Letters **90**, 104101 (2007).

2 Protagonists and Antagonists

The main players in this thesis are mechanical oscillators and ultracold atoms. Before we study the interaction of these, the theoretical foundations are briefly reviewed in this chapter.

2.1 Mechanical oscillators

A massive structure which is subjected to boundary conditions can perform mechanical oscillations. Intrinsic elasticity of the structure provides a restoring force towards the equilibrium position and gives rise to oscillations about this equilibrium. Free oscillations are a superposition of orthogonal eigenmodes, which are each characterized by an eigenfrequency and a modefunction, the latter describes the geometric shape of the structure for maximum displacement from the equilibrium. In our experiments where the motion of such mechanical oscillators is coupled to ultracold atoms we always make use of the fundamental mechanical mode, which is the mode with the lowest eigenfrequency. Fig. 2.1 shows a variety of mechanical oscillators, with sizes ranging from the centimeter range down to nanometers. In this thesis, we employ a silicon nitride membrane as an end mirror in chapter 4, an AFM cantilever in chapter 3, and a mechanical oscillator with a 2×2 μm^2 mirror in chapter 5.

Mechanical oscillators are employed in a series of recent experiments, and form an active field of research. Optomechanical experiments [6, 7, 8, 9] are on the way to achieve ground state occupation of a single mode of a mechanical oscillator. New transduction schemes of nanomechanical motion allow to drive and read out mechanical oscillators without the need of functionalization with gates, magnets or mirros [52]. Other applications are to detect single electron spins [5], or to establish an alternative (electric) current standard [53].

2.1.1 Analytical description

In this section, I give an analytical description of mechanical oscillators, and focus on the description of the mechanical oscillators employed in this thesis. The type of mechanical oscillator used in chapters 3 and 5 is a single-side clamped, cantilever oscillator, and a membrane oscillator is employed in chapter 4, see Fig. 2.2 for an overview.

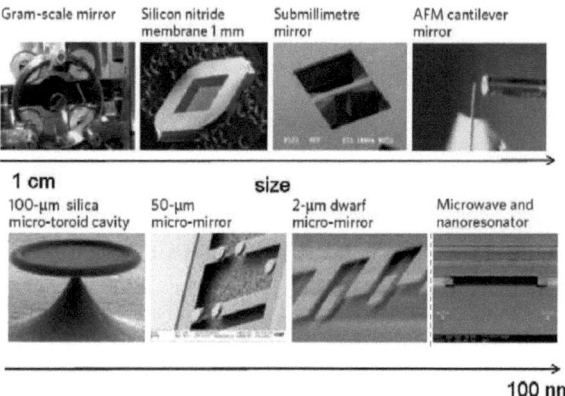

Figure 2.1: Mechanical oscillators of various sizes. A silicon nitride membrane is employed as an end mirror in chapter 4, an AFM cantilever is used in chapter 3 to couple to a BEC via the surface potential, and the mechanical oscillator with a 2×2 μm^2 mirror is characterized in chapter 5. Figure taken from [9].

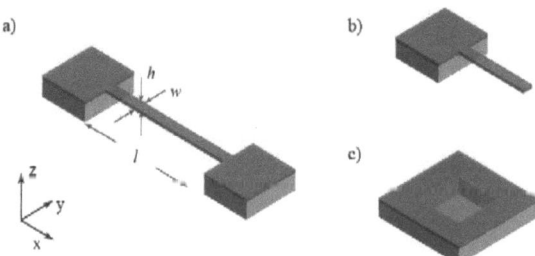

Figure 2.2: Typical mechanical oscillators. Single-side clamped mechanical oscillator are used in chapters 3 and 5, a membrane oscillator is employed in chapter 4 as an end mirror for an optical lattice. Figure taken from [54].

2.1 Mechanical oscillators

Cantilever oscillator – Euler-Bernoulli theory A mechanical cantilever oscillator consists of a massive beam, which is clamped to a support at one side, see Fig. 2.2b. The structure is micro- or nanofabricated, and the beam consists of an elastic material like silicon nitride (SiN) or silicon (Si). The dynamic properties are characterized by the mass density ρ and the modulus of elasticity E (Youngs Modulus). The mode functions and the eigenfrequencies can be explicitly calculated from the Euler-Bernoulli theory [55].

We consider a beam with length l, width w and height h, and assume that the length is large compared to width and height. The beam is supported at the small cross $A = wh$ section at one end ($z = 0$), and free at $z = l$. The transverse displacement $U(z,t)$ is described by

$$EI_y \frac{\partial^4 U(z,t)}{\partial z^4} = -\rho A \frac{\partial^2 U(z,t)}{\partial t^2}, \tag{2.1}$$

with the bending moment $I_y = wh^3/12$, which is assumed to be constant along the beam. With the additional assumption of a harmonic time dependence of the displacement, $U(z,t) = U(z)e^{-i\omega t}$ with angular eigenfrequency ω. The spatial 'shape' of the mode function $u(z)$ is determined by

$$EI_y \frac{\partial^4 u(z)}{\partial z^4} = \rho A \omega^2 u(z), \tag{2.2}$$

and takes the general form

$$u(z) = a\cos(\beta z) + b\sin(\beta z) + c\cosh(\beta z) + d\sinh(\beta z) \tag{2.3}$$

with $\beta = (\rho A/EI_y)^{1/4} \omega^{1/2}$. The coefficients a, b, c, d are fixed by the boundary conditions

$$u(z=0) = \frac{\partial u(z=0)}{\partial z} = \frac{\partial^2 u(z=l)}{\partial z^2} = \frac{\partial^3 u(z=l)}{\partial z^3} = 0. \tag{2.4}$$

The 3rd and 4th conditions account for the requirement that there is zero transverse force and zero torque acting onto the free end. Inserting the boundary conditions into equation 2.4 yields $a_n = -c_n$ and $b_n = -d_n$, whereby n indexes the eigenmodes, and $n = 1$ represents the fundamental eigenmode. Moreover, the eigenfrequencies of a single-side clamped beam are determined by the solutions of the equation

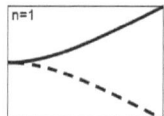 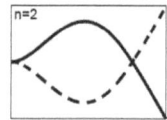 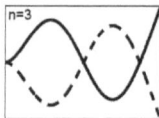

Figure 2.3: Modefunctions of the three lowest eigenmodes of a single side clamped beam oscillator. Figure taken from [54].

$$\cos \beta_n l \cosh \beta_n l + 1 = 0, \tag{2.5}$$

which are numerically determined to be $\beta_n l = (1.875, 4.694, 7.855...)$. The modefunction of a single-side clamped beam is obtained by inserting the above mentioned conditions for a_n, b_n, c_n, d_n and $\beta_n l$ into equation 2.4, and Fig. 2.3 shows the three lowest mode functions for $n = 1, 2, 3$ of such a single-side clamped beam. The angular eigenfrequencies are given by

$$\omega_n = \sqrt{\frac{EI_y}{\rho A}} \beta_n^2, \tag{2.6}$$

which can be simplified to

$$\omega_1 = 2\pi \times 0.1615 \sqrt{\frac{E}{\rho}} \frac{h}{l^2} \tag{2.7}$$

for the fundamental mode frequency.

In chapter 3 we employ a 'typical' micromechanical (AFM) cantilever. Such cantilevers have frequences in the range $\omega_1 = 10..500$ kHz$/2\pi$, and typically mechanical quality factors of $Q = 10^3..10^5$. In chapter 5, we study nanomechanical oscillators which have a typical length of a few microns, eigenfrequencies in the range $\omega_1 = 1..100$ MHz$/2\pi$, and typically mechanical quality factors of $Q = 10^3..10^5$ [56].

Membrane oscillator The Euler-Bernoulli theory describes mechanical oscillators where only the intrinsic elasticity gives rise to a restoring force. The situation is more complicated as soon as stress or strain contributes to the restoring force. The mechanical oscillator used in chapter 4 is a thin SiN membrane which is stretched

2.1 Mechanical oscillators

onto a rectangular frame. The frame imposes a uniform tension S, which is perpendicular to the membrane edges and directed outwards. If the restoring force due to stress is large compared with the restoring force due to elasticity, the role of elasticity can be neglected in the description of the membrane. We show in section 4.3.1, that the behaviour of the membranes used in our experiment can be well described by neglecting the contribution of elasticity to the restoring force.

Fig. 2.2c shows a sketch of a quadratic silicon nitride membrane (SiN). Such mechanical oscillators are fabricated by deposition of a SiN layer on top of a silicon wafer substrate in a LPCVD process. The tensile stress of the SiN layer can be explained from a 'lattice' mismatch of the amorph SiN layer and the underlying silicon wafer. After deposition of the SiN layer, an anisotropic KOH etch is employed to remove the silicon wafer selectively, leaving the SiN membrane freestanding [57]. The stress can be engineered by adjusting the LPCVD process parameters, and the typical tensile stress of low stress SiN membranes[1] used in our experiments or in [58], is around 140 MPa. Quality factors of up to 1.5×10^6 are achieved in our room temperature setup at an eigenfrequency of $\omega_{11} = 270$ kHz. In principle, the tensile stress of the SiN layer can be increased, and values around 1.4 GPa are favourable for realizing double-side clamped string oscillators with mechanical quality factors of up to $Q = 2 \times 10^5$ at several tens of MHz [59, 52].

With the assumption that the restoring force is only due to the uniform tension S as described above, all possible modefunctions of a rectangular membrane are given by the Fourier series [60]

$$w(x,y) = \sum_{i=1}^{\infty} \sum_{j=1}^{\infty} \Phi_{ij} \sin\frac{i\pi x}{a} \sin\frac{j\pi y}{b}, \qquad (2.8)$$

with the length of the edges a, b, height h and integer mode indices i and j. We apply the Lagrange formalism to derive an expression for the eigenfrequencies. The potential energy of the deflected membrane can be calculated from the area change due to deflection

$$\delta A = \int\int \sqrt{1 + \left(\frac{\partial w}{\partial x}\right)^2 + \left(\frac{\partial w}{\partial y}\right)^2}\, dxdy \qquad (2.9)$$

which transforms to

$$\delta A \approx \int\int 1 + \frac{1}{2}\left(\frac{\partial w}{\partial x}\right)^2 + \frac{1}{2}\left(\frac{\partial w}{\partial y}\right)^2 dxdy \qquad (2.10)$$

---
[1] ...which can be purchased at e.g. www.norcada.com

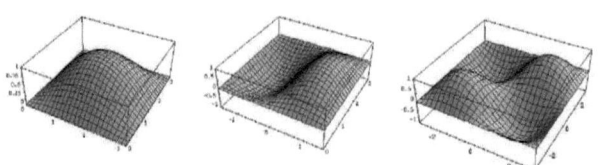

Figure 2.4: Modefunctions of the fundamental mode $(i,j) = (1,1)$ (left), the degenerate $(2,1)$ or $(1,2)$ mode, and the $(2,2)$ mode of a membrane oscillator.

for small deflections. The potential energy is given by

$$\Delta U \approx \frac{Sh}{2} \int\int \left[1 + \frac{1}{2}\left(\frac{\partial w}{\partial x}\right)^2 + \frac{1}{2}\left(\frac{\partial w}{\partial y}\right)^2\right] dxdy, \qquad (2.11)$$

and with the expression for the kinetic energy

$$T = \frac{\rho h}{2} \int\int \dot{w}^2 dxdy, \qquad (2.12)$$

the Lagrange formalism provides the equation of motion for free vibrations

$$\ddot{\Phi}_{ij} + \frac{S\pi^2}{\rho}\left(\frac{i^2}{a^2} + \frac{j^2}{b^2}\right)\Phi_{ij} = 0, \qquad (2.13)$$

with the height h of the membrane. Fig. 2.4 shows the mode functions of the lowest out-of-plane modes of a quadratic membrane. The eigenfrequencies are given by

$$\omega_{ij} = \sqrt{\frac{S\pi^2}{\rho}\left(\frac{i^2}{a^2} + \frac{j^2}{b^2}\right)}, \qquad (2.14)$$

which reduces for the fundamental mode $(i,j) = (1,1)$ and $a = b$ to

$$\omega_{11} = 2\pi \times \frac{1}{a}\sqrt{\frac{S}{2\rho}}. \qquad (2.15)$$

2.1 Mechanical oscillators

Effective mass The potential energy of a harmonically bound massive object can be expressed as $V = \frac{1}{2}m\omega^2 x^2$. This assumes that the massive object is point-like, and that whole mass is concentrated at the point which is displaced by x from the potential minimum. However, this is not the case for the mechanical oscillators that we have considered above where the moving mass is distributed over the whole mode function. The concept of the effective mass m_{eff} is introduced for describing such mechanical oscillators within the simple model. In order to calculate m_{eff}, a volume element is weighted according to the modefunction.

For a beam oscillator with uniform cross section and density along the beam length L, the effective mass m_{eff} is given by the 1D integration over the mode function along the beam [61]

$$m^i_{eff} = \frac{m}{L}\frac{\int_L u_i(y)^2 dy}{\int_L u_i(y) dy}, \qquad (2.16)$$

and gives $m_{eff} = (33/140)m$ for the fundamental mode $i = 1$ of a single-side clamped beam. For a membrane oscillator, an analogue 2D integral has to be solved, and the effective mass of a membrane oscillator as discussed above is given by $m_{eff} = (1/4)m$. The expression for the effective mass m_{eff} assumes that the amplitude x of the fundamental out of plane mode is measured at the position where the x attains a maximum, i.e. at the tip of a cantilever mechanical oscillator or in the center of a membrane oscillator.

2.1.2 Excitation, damping and thermal motion

Eigenmodes of mechanical oscillators can be excited in several ways. An eigenmode can be driven coherently with resonant excitation, as we use it in chapters 3 and 4. There, a piezo is driven at the eigenfrequency of an eigenmode, and radiates acoustic waves into the structure which sustains the mechanical oscillator. This leads to a resonant excitation of an eigenmode of the mechanical oscillator [53, 52]. The mechanical oscillators' motion can also be excited or damped by the modulation of the radiation pressure that a laser fields exerts onto the oscillator [62].

Another excitation mechanism which is always present is the coupling of the mechanical oscillator to a thermal bath, which excites the oscillator to thermal motion. On the other hand, the coupling leads to dissipation of energy in the oscillator, and to damping of the thermal motion.

Protagonists and Antagonists

Dissipation in mechanical systems

This section discusses selected sources of dissipation in mechanical oscillators under vacuum conditions[2]. The mechanisms which lead to mechanical dissipation are subject of current research, and not fully understood yet. The general observation is that the mechanical quality factor decreases with decreasing dimensions of the mechanical oscillator [64, 56]. A strong dependance of the mechanical quality factor on the surface to volume ratio is reported in [65].

Thermo-elastic dissipation Mechanical deformation of a solid introduces a variation of the strain field. This mechanism couples to phonons in the solid and transfers energy from the mechanical mode of interest to the local phonon field in the solid. The description of this process is based onto a thermal expansion coefficient which relates a length change of the solid to a temperature change [66], given that the timescale of the mechanical motion exceeds the timescale of thermalization. Dissipation is present if the thermal expansion coefficient is non-zero, and energy is transduced from mechanical motion into heat. This can be the dominant dissipation mechanism of mechanical oscillators under tensile stress, such as SiN membranes [67] or SiN strings [59].

Clamping losses Clamping of a mechanical oscillator gives rise to an energy transfer from the oscillator into the support. In practical implementations, the support is stiff, but still has a finite Youngs Modulus. In [68], a model is derived which predicts increased damping with the increase of the clamping area of an Euler-Bernoulli type mechanical oscillator. Similar work investigates the transmission of vibrational waves between joined, thin elastic plates of different widths [69] or the phonon tunneling from a mechanical oscillator into the support [70], finding an increase of the dissipation with the clamping area.

Coupling to local defects Depending on the temperature, local defects in or at the surface of bulk material can have a high impact on the damping. Local defects can be atoms or molecular compounds which are bound in an effective double-well potential [71]. At high temperatures, the double well structure does not play a role as the potential barrier height is much smaller than $k_B T$. At small temperatures where $k_B T$ is smaller than the barrier height, mechanical deformations of the mechanical oscillator couple to a local defect in the material via the deformation of the

---

[2]Background gas collisions lead to an additional damping, and the background gas can be modelled as a viscous medium [63]. This contribution to the overall damping rate is dominant for mechanical oscillators with a large surface at ambient atmosphere, but negligible under high vacuum conditions as provided in our experiments

2.1 Mechanical oscillators

double well potential. This leads to increased dissipation in the mechanical oscillator for decreasing temperature as observed in [72]. In [73, 67], a strong influence of tensile stress and stoichiometric composition onto the mechanical quality factor is reported. The density of local defects at the surface can be reduced with thermal annealing [74].

The individual sources of dissipation add up to the total dissipation of a mechanical oscillator, and are effectively described with the mechanical quality factor Q. Q^{-1} is proportional to the rate at which energy is dissipated in a mechanical oscillator. Q is a measure of the number of oscillation cycles that it takes the energy which is stored in a mechanical oscillator to decay to a fraction $1/e$ of the initial value in a ringdown measurement. In terms of the amplitude decay time τ, the mechanical quality factor is given by

$$Q = \frac{\omega \tau}{2} \qquad (2.17)$$

If the oscillation frequency is stable, i.e. a drift of the frequency due to e.g. temperature drifts is small compared to the measurement time, and for a not too high mechanical quality factor, Q can also be determined from a direct measurement in the frequency domain. With the FWHM κ of a Lorentzian fitted to the amplitude spectrum,

$$Q = \frac{\omega}{2\kappa}. \qquad (2.18)$$

In the measurements shown in chapter 4, the interaction of ultracold atoms with a membrane oscillator influences the damping of the membrane, and we observe this effect in ringdown measurements.

Thermal motion   The fluctuation-dissipation theorem states that systems that dissipate energy –as discussed above for mechanical oscillators– are also subject to noise. An uncorrelated Langevin noise force acts onto each point of the mechanical oscillator, and enforces thermalization with the thermal bath such that the equipartition theorem is fulfilled, i.e. each degree of freedom contains the same energy of $k_B T/2$ in thermal equilibrium. This yields

$$m_{eff}\omega^2 x_{rms}^2/2 = k_B T/2 \qquad (2.19)$$

for a harmonic mechanical oscillator, where x_{rms} is the rms amplitude of the thermal motion. Equivalently, each mechanical mode is occupied by

$$n = \frac{1}{e^{\hbar\omega_m/k_BT} - 1} \qquad (2.20)$$

phonons. The general solution for the differential equation of a harmonic, damped oscillator in the frequency domain is a Lorentzian. This is reflected in the equation for the spectral density of x_{rms}^2 [55, 75]

$$S(\omega) = \frac{S_F(\omega)}{m_{eff}^2} \frac{1}{(\omega_0^2 - \omega^2)^2 + (\omega_0^2/Q)^2}, \qquad (2.21)$$

with the spectrum of the statistic Langevin noise force $S_F(\omega) = 2k_B T m_{eff} \omega_0 / \pi Q$. The rms value of the thermal amplitude x_{rms} is given by

$$x_{rms}^2 = \int S(\omega) d\omega \approx \frac{k_B T}{m\omega^2}. \qquad (2.22)$$

This result is used to calibrate the sensitivity of a Michelson interferometer in section 4.2, where one of the two interferometer end mirrors is a membrane oscillator. The output of the interferometer is Fourier transformed, and the thermal motion shows up as a Lorentzian in the frequency domain. For calibration of the interferometer, the mean squared amplitude x_{rms}^2 equals the area below the Lorentzian, and a scaling factor is extracted which allows to determine the membrane amplitude in meters.

2.1.3 Towards quantum mechanical oscillators

The eigenmodes of a mechanical oscillator at room temperature are excited by the Langevin noise force to a thermal motion amplitude. Mechanical oscillators which are coupled to a thermal bath at room temperature behave classically due to their high phonon occupation. This would change for a phonon occupation number smaller than one, where mechanical oscillators enter the quantum ground state. Many groups investigate optomechanical cooling of the thermal motion of a mechanical oscillator experimentally, for a review see [76, 8, 9] with a technique which is similar to laser cooling of atoms [15, 16]. The displacement of a mechanical oscillator is coupled to the light field inside an optical cavity, and the motion of the oscillator detunes the cavity resonance with respect to the frequency of the light field which leads to a change of the intracavity power and tunes the radiation pressure of the light field onto the oscillator. This coupling is exploited to reduce the initial phonon number occupation of a single mechanical mode which is in most cases the fundamental center of mass mode. This method can be applied to relatively large mechanical

2.1 Mechanical oscillators

oscillators which can reflect laser beams, e.g. SiN membranes [58], whereas mechanical oscillators with sub-wavelength dimensions can be cooled in a cryostat where the phonon number occupation of all modes is simultaneously reduced [12, 13, 14]. An advantage of sub-wavelength sized oscillators in the respective of reaching the quantum ground state is their relatively high eigenfrequency. In order to cool the membrane employed in the experiments reported in [58] with an eigenfrequency of $\omega/2\pi = 130$ kHz to the quantum ground state cryogenically, $T \ll 6.5$ μK would have to be reached. In comparison, [14] studies a mechanical oscillator with an eigenfrequency of $\omega/2\pi = 6$ GHz, which attains the quantum groundstate already at $T \simeq 100$ mK.

In the following, I will give a short overview of applications of mechanical oscillators.

Sensor applications Cooling of the mechanical mode which is used for sensing could increase the sensitivity of mechanical oscillators with respect to forces [1, 2], temperature changes [3] and additional masses [4]. Magnetic resonance force microscopy allows to detect a single electron spin in a solid [5]. Interferometric gravitational wave detection would also benefit from a reduced thermal amplitude of the interferometer end mirrors [77]. The preferred cooling mechanism for these applications is cryogenic, since it achieves a permanent reduction of the phonon number occupation, whereas the cooling in optomechanical schemes is only present as long as the laser is switched on.

Gravity in quantum mechanics The ground state of a mechanical oscillator which is coupled to a thermal bath decoheres at a rate $\gamma \propto \kappa n_{th} = (\omega/2Q)n_{th}$. The source of decoherence in quantum mechanics is coupling to the environment. However, there are non-standard theories which predict additional, intrinsic decoherence of a quantum state due to gravity. In [78], the self energy E_Δ of the difference of the mass distributions, which represent the two states of a superposition state, is considered to give rise to an intrinsic decoherence. The lifetime T of such a superposition is found to be limited by an additional decoherence rate $T \propto \hbar/E_\Delta$. This effects could be observed [79] in an experiment with a ground-state mechanical oscillator in a superposition state.

Hybrid quantum systems A potential application for mechanical oscillators in the ground state is the realization of 'hybrid quantum systems' which comprise two quantum systems of different nature or origin, like e.g. ground state mechanical oscillators and a two-level quantum system. Non-classical states could be prepared by coupling a mechanical oscillator to a well-controlled two-level system. Extending the already demonstrated quantum control of atoms [18, 19, 20, 21], ions [22] or superconducting flux qubits [23, 24] onto mechanical oscillators could allow for the preparation of non-classical quantum states in a mechanical mode.

One approach is the coupling of mechanical oscillators to solid-state quantum systems [10, 11, 12, 13]. These are strongly coupled to the environment which results in relatively short coherence times, but also in fast coupling rates to mechanical oscillators which are fabricated nearby on the same chip. The experiments can be operated without optical access which facilitates experiments at cryogenic temperatures. In addition, the cryogenic temperatures cool the mechanical oscillator to low phonon occupation numbers. The effect of a solid state qubit system onto the mechanical oscillator is already demonstrated. In [80], dispersive coupling of a superconducting qubit to the center of mass mode of a mechanical oscillator is reported. Recently, a dilatational volume mode of a mechanical oscillator at a frequency of 6 GHz was cooled to the ground state in a cryogenic environment, and coupled to a Josephson flux qubit. The control on the single-phonon level which has been achieved in this hybrid quantum system is reported in [14].

A second approach is to couple mechanical oscillators to atomic quantum systems. A variety of theoretical proposals [27, 28, 29, 30, 31, 32, 33, 34, 35, 36, 37, 38, 39, 40] suggests the experimental realization of such systems, involving e.g. neutral atoms [31, 38, 39, 30, 33, 34, 35, 36, 37]. Atomic systems studied are very well isolated from sources of decoherence which allows one to achieve long coherence times [47, 26]. Despite of the numerous proposals, only few experiments [41, 81, 82] have investigated the coupling of mechanical oscillators and ultracold atoms experimentally. In chapters 3 and 4, I describe experiments [81, 82] that have been pursued in this thesis.

2.2 Ultracold atoms

In this thesis, I report on experiments which realize a controlled coupling of microstructured mechanical oscillators to ultracold neutral atoms. Ultracold atoms are very well isolated from the environment [26], and all degrees of freedom can be controlled on a quantum level [44, 21, 46]. Atoms in trapping potentials can be regarded as mechanical oscillators. In the experiments reported in chapters 3 and 4, the vibrations of mechanical oscillators are coupled to the center of mass mode (COM) of atoms in a trapping potential. In chapter 5 we consider the coupling of a mechanical oscillator to the collective spin of a magnetically trapped Bose-Einstein condensate (BEC).

2.2.1 Magnetic trapping on atoms chips

Magnetic trapping of neutral atoms relies on the interaction of the magnetic moment $\boldsymbol{\mu}$ of an atom with an external magnetic field $\boldsymbol{B}$ [49]. In a classical model, the interaction of the magnetic moment $\boldsymbol{\mu}$ with an external magnetic field $\boldsymbol{B}(\boldsymbol{r})$ is decribed by

$$V(\boldsymbol{r}) = -\boldsymbol{\mu} \cdot \boldsymbol{B}(\boldsymbol{r}) = -\mu B(\boldsymbol{r}) \cos \theta, \qquad (2.23)$$

and can be interpreted as a precession of the magnetic moment about the magnetic field axis, with constant angle θ between $\boldsymbol{\mu}$ and $\boldsymbol{B}$. The constant angle θ requires a quantum system to remain in the initial quantum state, as discussed below in the quantum modell of the atom-field interaction.

The behaviour of alkali atoms is dominated by the valence electron [83]. The orbital angular momentum $\boldsymbol{L}$ of the outer electron and its spin angular momentum $\boldsymbol{S}$ are coupled to the total angular momentum $\boldsymbol{J} = \boldsymbol{L} + \boldsymbol{S}$ which results in the fine structure splitting and gives rise to the D_1- and D_2-line. The total angular momentum of the electron $\boldsymbol{J}$ couples in turn to the angular momentum of the nucleus $\boldsymbol{I}$ which results in the total angular momentum of an atom $\boldsymbol{F} = \boldsymbol{I} + \boldsymbol{J}$. This coupling causes a hyperfine splitting of each level, e.g. the ground state $5^2S_{1/2}$ of ^{87}Rb is split into two hyperfine levels, and the fine splitted branch $5^2P_{3/2}$ (D_2 line) into four excited states. Each of these hyperfine levels contains $2F + 1$ magnetic sublevels, which are degenerate if no magnetic field is applied. The degeneracy is lifted for $\boldsymbol{B} \neq 0$, and the energy levels are given by

$$E(m_F) = \mu_B g_F m_F B(\boldsymbol{r}) = V(\boldsymbol{r}), \qquad (2.24)$$

with the quantum number m_F which is associated with the component of $\boldsymbol{F}$ along the direction of the magnetic field $\boldsymbol{B}(\boldsymbol{r})$, the g-factor g_F and the Bohr magneton μ_B. Comparison of the classical and quantum mechanical picture shows that $\cos \theta$ corresponds to m_F/F in a geometric interpretation.

The sign of $g_F m_F$ determines the direction of the force which is experienced by a particle in a magnetic field. For $g_F m_F > 0$, the particle is attracted towards the minima of the magnetic field (weak-field-seeking state). $g_F m_F < 0$ yields attraction to magnetic field maxima. According to Maxwell's equations, magnetic field maxima can not be realized in free space and we focus the description onto the confinement of weak-field-seeking states in magnetic field minima.

In order to achieve stable magnetic trapping, the atom should not change its quantum state, i.e. the quantum number m_F should not change. However, this can occur if the atomic spin does not follow the direction of the magnetic field adiabatically for the motion of an atom in the trapping potential. Adiabaticity is fulfilled if the change of the magnetic field in the reference frame of an atom is slower than the (Larmor) precession frequency ω_L of the atoms' magnetic moment in the external magnetic field

$$\frac{d\theta}{dt} < \frac{\boldsymbol{\mu}|\boldsymbol{B}(\boldsymbol{r})|}{\hbar} = \omega_L. \tag{2.25}$$

If this equation is not true, the state of the atom can change from an initially weak-field-seeking state to e.g. an untrapped motional state. The loss of atoms from the trapped ensemble is called Majorana loss, and can be reduced by increasing the minimum precession frequency ω_L in trap. This is done experimentally by choosing a relatively high value for the magnetic field at the minimum of the trapping potential.

Magnetic trapping of neutral atoms on an atom chip

Magnetic trapping potentials can be engineered with atom chip technology [84, 85, 86, 87, 88], where a magnetic near field emanates from an electric current which flows through a microfabricated wire. Magnetic field minima which are suitable for trapping of ultracold atoms in weak-field-seeking states can be provided by superimposing the near field of a chip wire with a homogeneous external field. Fig. 2.5 shows the general principle. Atom chips provide a robust toolbox to achieve [42, 43] and study or coherently manipulate [44, 21, 46] BEC in compact setups [89]. BECs can be trapped close to surfaces without compromising on coherence times [47]. Since the position of a magnetic trap is referenced to the wires on the atom chip, we can position a BEC can be reproducibly approached to the solid state surface, and in particular to stuructures like e.g. mechanical oscillators. The scalability of microfabrication allows to create elaborate potential landscapes [85]. The method proposed in [90] allows to automize the design microfabricated structure in order to realize complex potential landscapes.

Trapping potentials can be classified by the absolute value of the magnetic field at the trap minimum, which can be either zero or non-zero [85].

Quadrupole trap If the magnetic field minimum is zero, the trapping potential is quadrupole-like. It can be approximated by a linear function $\boldsymbol{B} = B'_x x \boldsymbol{e}_x + B'_y y \boldsymbol{e}_y + B'_z z \boldsymbol{e}_z$, with $B'_x + B'_y + B'_z = 0$. This configuration, which is conventionally realized with two coils in anti-Helmholtz configuration, can also be provided by the near field arising from a current through a U-shaped wire and a superimposed offset field. The

2.2 Ultracold atoms 21

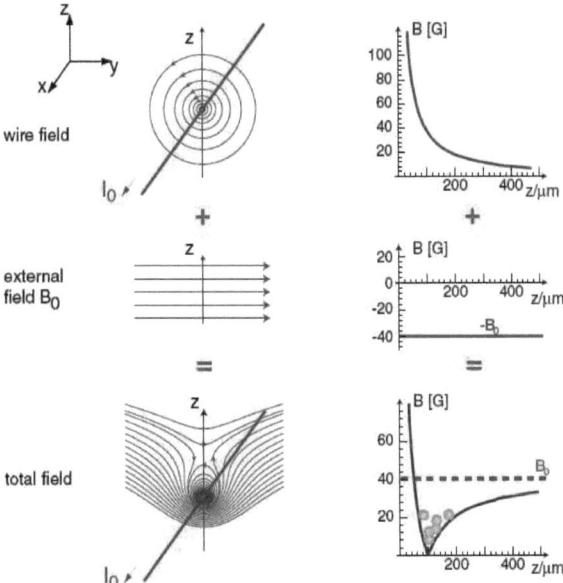

Figure 2.5: A 2D trapping potential can be created by superimposing the magnetic field of a current carrying wire and an external field. The example on the right-hand side assumes a wire current $I = 2$ A and an external field $B_{0\perp} = 4$ mT. The magnetic field gradient at the trap center is $|B'(r_0)| = 40$ T/m. Figure taken from [85].

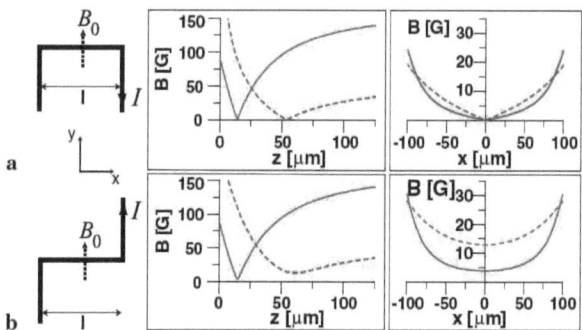

Figure 2.6: (a) Quadrupole trap and (b) IoffePritchard trap. The wire layout and external field direction is sketched on the left-hand side, the right hand side shows the magnetic potentials for $l = 250$ μm and $I = 2$ A. The external field along y is $B_0 = 5.4$ mT (dashed lines) and 16.2 mT. Figure taken from [85].

magnetic field due to the anti-parallel currents in the bent wires cancels at the trap center, such that the minimum of the magnetic field is zero. Fig. 2.6 (a) shows the wire geometry and the magnetic field. The zero magnetic field in the trap center gives rise to Majorana losses due to spin flip transitions to untrapped motional states. This occurs, when an atom travels through the magnetic field minimum where the quantization axis is not well-defined. The size of this effective trap 'hole' for an atom with velocity v is $\sqrt{2\hbar v/\pi\mu B'}$. This trapping geometry can be used for a relatively hot atomic cloud, where most of the atoms occupy orbits far away from the hole.

Ioffe-Pritchard trap Majorana spin flips can be reduced if a well-defined quantization axis exists throughover the trapping potential, which means that a zero magnetic field should be avoided. The resulting Ioffe-Pritchard type potential can be approximated to lowest order by a harmonic potential. The field configuration with a bias field $B_{0\parallel}$ along the x-axis is approximately

$$\boldsymbol{B} = B_{0\parallel} \begin{pmatrix} 1 \\ 0 \\ 0 \end{pmatrix} + B' \begin{pmatrix} 0 \\ -y \\ z \end{pmatrix} + \frac{B''}{2} \begin{pmatrix} x^2 - \tfrac{1}{2}(y^2 + z^2) \\ -xy \\ -xz \end{pmatrix}. \qquad (2.26)$$

Close to the minimum, an atom oscillates along the x-axis and the radial axes with the trap frequencies

2.2 Ultracold atoms

$$\omega_x = \sqrt{\frac{\mu_B g_F m_F}{m}} \sqrt{B''} \qquad (2.27)$$

$$\omega_\perp = \sqrt{\frac{\mu_B g_F m_F}{m} \left(\frac{B'^2}{B_0} - \frac{B''}{2} \right)}. \qquad (2.28)$$

Fig. 2.6 (b) shows the wire geometry and magnetic field of a chip based Ioffe-Pritchard trap. The contributions of the two wires, which are bent to opposite directions adds in the trap center to a non zero magnetic field. The external field is adjusted such that the magnetic field at the minimum is non-zero.

Trap configurations used in our experiment

In the experiment described in chapter 3 atomic ensembles are trapped, transported and manipulated on an atom chip. Besides a standard Ioffe-Pritchard trap which loads atoms from the mirror-MOT to the chip, we combine the basic trap geometries discussed above with additional fields from external coils and additional wires.

- **Transport of atoms in a waveguide** An application of a quadrupole trap is the transport of a relatively hot atomic ensemble over large distances [91]. A current carrying wire in combination with a homogeneous field provides a 2D confinement which is translational invariant along the axis of the wire. This waveguide potential is superimposed with an additional quadrupole field which is created by coils in anti Helmholtz configuration in order to confine the atomic ensemble in the waveguide. A shift of the quadrupole field minimum shifts the trap minimum along the waveguide. This is used to transport atoms over millimeter distances in the experiment described in chapter 3.

- **Modulation of the longitudinal waveguide potential** The translationally invariant waveguide potential can be modulated with a current through a wire which intersects the wire providing the waveguide perpendicularly, as illustrated in Fig. 2.7. This can be used to achieve a strong longitudinal confinement. We prepare Bose-Einstein condensates in a Ioffe-Pritchard type trap geometry. A current flow through this wire provides strong longitudinal confinement, and diggs a 'dimple' into the center of the trapping potential. Several parallel dimple wires can be used to shift the position of the magnetic trap along the central wire.

Bose-Einstein condensation (BEC) was achieved in atom chip based magnetic traps as discussed above roughly ten years ago [42, 43]. In setups following the design of [42], typically $10^3..10^4$ atoms are condensed at trap frequencies ranging from $\omega/2\pi = 1..15$ kHz [81]. The theory of BEC is briefly reviewed in 3.1.3.

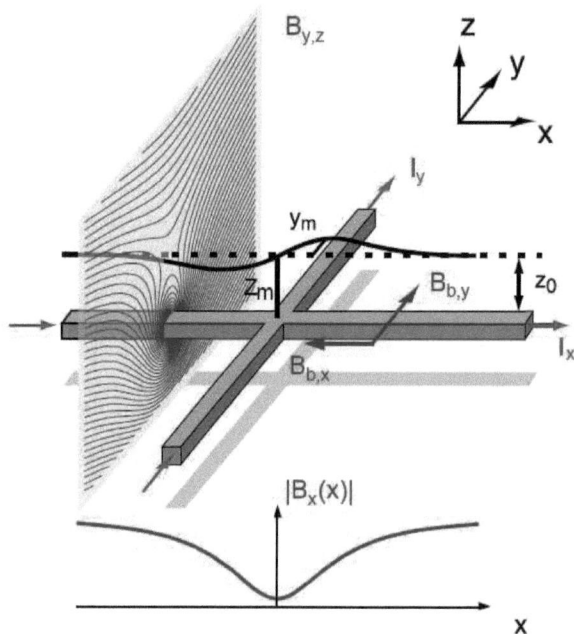

Figure 2.7: The intersection of two waveguides provides a barrier in a 2D waveguide. Figure taken from [54].

2.2 Ultracold atoms

2.2.2 Optical trapping

Optical potentials are a dynamic field of research, and one their manifold applications is to simulate Hamiltonians known from solid state physics, e.g. to study the phase transition from a superfluid to a mott insulator with bosons [92, 18] or to study ferromagnetism with fermions [93].

In the experiment described in chapter 4, we load ultracold ^{87}Rb atoms into a standing wave optical potential, and realize a physical interaction between optical potentials and solid state physics. This section reviews the interaction of an atom with a far detuned light field, see [94].

Neutral atoms in light fields An electric field $\boldsymbol{E}(\boldsymbol{r},t) = \hat{e}E(\boldsymbol{r})\exp(-i\omega t) + c.c.$ induces an oscillating dipole moment in an atom [94]. The complex amplitude d of the dipole moment is related to the field amplitude E [95] by the polarizability $\alpha(\omega)$ which depends on the laser frequency ω,

$$d = \alpha(\omega)E. \tag{2.29}$$

The time averaged interaction of the dipole moment and the electric field is described by the potential

$$V_{dip}(\boldsymbol{r}) = -\frac{1}{2}\langle \boldsymbol{dE}\rangle = -\frac{1}{2\epsilon_0 c}\Re(\alpha(\omega))I(\boldsymbol{r}) \tag{2.30}$$

with the intensity $I = 2\epsilon_0 c|E|^2$. The factor $1/2$ in equation 2.30 accounts for the fact that the dipole moment is not a permanent, but an induced moment. The dipole force onto an atom is given by the gradient of the interaction potential, and provides a conservative potential as long as photon scattering can be neglected. The photon scattering rate is given by

$$\Gamma_{sc}(\boldsymbol{r}) = \frac{\langle \boldsymbol{dE}\rangle}{\hbar\omega} = \frac{1}{\hbar\epsilon_0 c}\Im(\alpha(\omega))I(\boldsymbol{r}). \tag{2.31}$$

In order to determine $\alpha(\omega)$ one assumes that an electron is elastically bound to the nucleus of an atom. The situation can be modelled as a damped harmonic oscillator which is driven from a laser field which oscillates at the frequency ω. The damping is due to power radiation of the accelerated charge. Integration of the equation of motion gives the result

$$\alpha(\omega) = 6\pi\epsilon_0 c^3 \frac{\Gamma_{se}/\omega_0^2}{\omega_0^2 - \omega^2 - i(\omega^3/\omega_0^2)\Gamma_{se}}, \qquad (2.32)$$

with the eigenfrequency ω_0 of the unperturbed atom without light field and the damping rate Γ_{se}.

In the following, we show how the atomic polarizability $\alpha(\omega)$ can be determined from a semiclassical approach where the atom is modelled as a two-level system which interacts with the light field. As long as saturation effects can be neglected, the result corresponds to the result derived from the oscillator model. However, the damping rate Γ_{se} is determined from the dipole matrix element between ground and excited state for the dipole operator $\boldsymbol{\mu} = -e\boldsymbol{r}$

$$\Gamma_{se} = \frac{\omega_0^3}{3\pi\epsilon_0 \hbar c^3}|\langle e|\boldsymbol{\mu}|g\rangle|^2. \qquad (2.33)$$

The condition of negligible saturation is usually fulfilled for dispersive trapping of atoms, with large detuning $\Delta = \omega - \omega_0$. The expression for the polarizability α allows us to give explicit expressions for the dipole potential $V_{dip}(\boldsymbol{r})$ and the scattering rate $\Gamma_{sc}(\boldsymbol{r})$

$$V_{dip}(\boldsymbol{r}) = -\frac{3\pi c^2}{2\omega_0^3}\left(\frac{\Gamma_{se}}{\omega_0 - \omega} + \frac{\Gamma_{se}}{\omega_0 + \omega}\right)I(\boldsymbol{r}) \approx \frac{3\pi c^2}{2\omega_0^3}\frac{\Gamma_{se}}{\Delta}I(\boldsymbol{r}) \qquad (2.34)$$

$$\Gamma_{sc}(\boldsymbol{r}) = \frac{3\pi c^2}{2\hbar\omega_0^3}\left(\frac{\omega}{\omega_0}\right)^3\left(\frac{\Gamma_{se}}{\omega_0 - \omega} + \frac{\Gamma_{se}}{\omega_0 + \omega}\right)^2 I(\boldsymbol{r}) \approx \frac{3\pi c^2}{2\hbar\omega_0^3}\left(\frac{\Gamma_{se}}{\Delta}\right)^2 I(\boldsymbol{r}) \qquad (2.35)$$

The counter-rotating term at $\omega + \omega_0$ is neglected within the rotating-wave approximation ($|\Delta| \ll \omega_0$). The scattering rate Γ_{sc} can also be expressed in terms of the dipole potential

$$\hbar\Gamma_{sc} = \frac{\Gamma_{se}}{\Delta}V_{dip}, \qquad (2.36)$$

which links the absorptive and dispersive response of the atom to the light field. For 'red' detuning $\Delta < 0$, the dipole potential is negative and the atom is attracted towards the intensity maxima of the light field. This can be employed to realize trapping potentials for laser-cooled atoms. The scaling of $V_{dip}(\boldsymbol{r}) \propto I/\Delta$ and $\Gamma_{sc} \propto I/\Delta^2$ suggests to choose large detuning and laser power in order to minimize the scattering rate.

2.2 Ultracold atoms

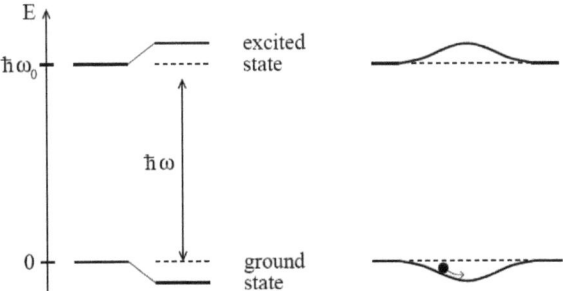

Figure 2.8: The interaction with the red detuned light field shifts the energy levels of ground and excited states of a two-level atom. Figure taken from [94].

Multi-level atoms We consider a multi-level atom with a ground and excited state in order to get a deeper insight into the origin of the trapping potential. In second order perturbation theory, a far detuned light field shifts the energy of the i-th state with unperturbed energy $\mathcal{E}_i$ by

$$\Delta E_i = \sum_{j \neq i} \frac{|\langle j|H_1|i\rangle|^2}{\mathcal{E}_i - \mathcal{E}_j}, \tag{2.37}$$

with interaction Hamiltonian $H_1 = -\boldsymbol{\mu}\boldsymbol{E}$ and the electric dipole operator $\boldsymbol{\mu} = -e\boldsymbol{r}$. For a two-level atom with $H_1 = -\boldsymbol{\mu}\boldsymbol{E}$, one finds

$$\Delta E = \pm \frac{|\langle e|\boldsymbol{\mu}|g\rangle|^2}{\Delta}|E|^2 = \pm \frac{3\pi c^2}{2\omega_0^3}\frac{\Gamma_{se}}{\Delta}I, \tag{2.38}$$

where the negative (positive) sign designates the ground (excited) state energy shift. The situation is shown in Fig. 2.8. The energy shift of an atom due to the light field is the origin of the optical dipole potential in the case of low saturation where atoms reside mainly the ground state.

In order to derive the dipole potential for a multi-level alkali atom, one has to find an expression for the transition matrix elements μ_{ij} between the ground state $|g_i\rangle$ and the excited state $|e_j\rangle$. The matrix elements μ_{ij} can be expressed by means of a reduced matrix element $\|\mu\|$ which is related to the spontaneous decay rate 2.33. The real transition coefficients c_{ij} account for the coupling strength between the sub-levels i and j. The matrix element is given by $\mu_{ij} = c_{ij}\|\mu\|$, with the energy shift of the ground state

$$\Delta E_i(\mathbf{r}) = \frac{3\pi c^2 \Gamma_{se}}{2\omega_0^3} I(\mathbf{r}) \times \sum_j \frac{c_{ij}^2}{\Delta_{ij}}. \tag{2.39}$$

In our experiment, we use alkali atoms (^{87}Rb) with a fine structure which is split into a D_1 and D_2 line $^2S_{1/2} \to {}^2P_{1/2}, {}^2P_{3/2}$ at 795nm and 780nm, respectively. For large detuning Δ of the light field with respect to the optical transition, the hyperfine structure splitting is not resolved and the dipole potential for an atom with total angular momentum F and magnetic quantum number m_F can be written as

$$V_{dip}(\mathbf{r}) = \frac{\pi c^2 \Gamma_{se}}{2\omega_0^3} \left(\frac{2 + \mathcal{P}g_F m_F}{\Delta_{2,F}} + \frac{1 - \mathcal{P}g_F m_F}{\Delta_{1,F}} \right) I(\mathbf{r}), \tag{2.40}$$

with g-factor g_F, and $\mathcal{P} = 0, \pm 1$ for linearly and circularly polarized $\sigma^\pm$ light. This equation holds as long as the detunings $\Delta_{1,F}, \Delta_{2,F}$ of the light field with respect to the center of the hyperfine split excited states of the D_1 and D_2 line are large compared to the excited-state hyperfine splitting. The hyperfine splittings of ^{87}Rb are of the order of a few hundred MHz.

For linearly polarized light the dipole potential due to the total light intensity $I(r)$ is independent of m_F and can be written as

$$V_{dip}(\mathbf{r}) = \frac{\hbar \Gamma_{se}^2}{8} \frac{1}{\Delta} \frac{I(\mathbf{r})}{I_{sat}}, \tag{2.41}$$

with the saturation intensity $I_{sat} = \hbar \Gamma_{se} \omega_0^3 / 12\pi c^2$ and the detuning

$$\frac{1}{\Delta} = \frac{1}{3} \left(\frac{1}{\Delta_{1,F}} + \frac{2}{\Delta_{2,F}} \right). \tag{2.42}$$

The scattering rate in this situation is given by

$$\Gamma_{sc} = \frac{\pi c^2 \Gamma_{se}^2}{2\hbar \omega_0^3} \left(\frac{2}{\Delta_{2,F}^2} + \frac{1}{\Delta_{1,F}^2} \right) I(\mathbf{r}). \tag{2.43}$$

2.2 Ultracold atoms

1D lattice potential In the experiment described in chapter 4 a red detuned gaussian laser beam is reflected at a mirror such that the incoming and the outbound beam are overlapped. The modulation depth of the resulting interference pattern is given by

$$I = 4I_0|L| \left(\frac{w_0}{w(z,z_0)}\right)^2 e^{-\frac{2r^2}{w(z,z_0)^2}} \cos\left(kz + \frac{kr^2}{2R(z,z_0)}\right)^2, \qquad (2.44)$$

with wave vector k of the incoming laser beam which is parallel to the z-axis. The factor $|L| = |R||T|^2$ accounts for the typical experimental situation, that the reflected beam might be weaker than the incoming beam due to the amplitude reflectivity $|R|$ of the mirror, and the amplitude transmittivity $|T|$ of the optical components in the beam path between the atoms and the mirror. The laser beam of power P is focussed to a waist[3] w_0 at the position z_0 on the z-axis. The peak intensity of a single beam is given by $I_0 = 2P/\pi w_0^2$. The waist of the gaussian beam along the z-axis is given by $w(z,z_0) = w_0\sqrt{1 + (z/z_{ray})^2}$. The radius $w(z,z_0)$ increases within the Rayleigh length $z_{ray} = \pi w_0^2/\lambda$ from the waist position at z_0 by a factor $\sqrt{2}$. The curvature of the wavefronts due to the beam divergence is given by the radius of curvature $R(z,z_0) = z(1 + (z_{ray}/z)^2)$.

Atoms trapped in such a lattice potential are longitudinally confined in a $\cos^2$ potential, and this potential can be approximated by a harmonic potential for cold ensembles with $k_B T \ll \max(V_{dip})$. In this limit, the same approximation can be used for the axial confinement which is given by the gaussian beam profile. The trap frequencies of the harmonic approximation are given[4] [96] by

$$\omega_{ax} = 2\pi\sqrt{\frac{2V_{dip}}{m\lambda^2}} \qquad (2.45)$$

$$\omega_{rad} = \sqrt{\frac{4V_{dip}}{mw_0^2}} \qquad (2.46)$$

In the experimental situation that we encounter in chapter 4, the temperature of the ensemble is of the order of the trap depth ($k_B T \approx \max(V_{dip})$), and the harmonic approximation is no longer valid. The spectrum of a trapped ensemble is further broadened due to a variation of the intensity in the lattice due to the nature of the gaussian beam. The peak intensity of lattice sites for increasing distance from the beam waist position decreases due to the divergence of the gaussian beam and leads

---

[3] w_0 is the waist of the beam, which is defined as the shortest distance from the intensity maximum to the point, where the intensity is decreased to $1/e^2$.
[4] Note, that V_{dip} is composed of the modulation depth given by 2.44 and an offset

to a spread of trap frequencies along the lattice. The strength of this effect depends on the relation between the Rayleigh length and the size of the occupied lattice. If the latter is small compared to the Rayleigh length, this axial spread of trap frequencies can be small. In addition to this spread of trap frequencies along the lattice, the intensity of the laser beam is also modulated transversally. Depending on the temperature of the ensemble, atoms can orbit in regions with $I \ll I_0$ which causes a transverse spread of trap frequencies. This effect can be considerable, if the temperature of the atoms is of the order of the potential depth. These effects are discussed quantitatively in section 4.3.3

The center of mass motion of an atomic ensemble in the optical lattice leads to a redistribution of photons between the travelling waves which form the optical lattice, and effectively to a power modulation of a laser beam. The resulting modulation of the light pressure onto the mechanical oscillator couples the atomic center of mass motion to the motion of the oscillator. This mechanism is discussed in 4.1.2.

Laser cooling in optical lattices In the experiment described in chapter 4, atoms trapped in an optical potential are coupled to a mechanical oscillator. The objective is to transfer energy from the mechanical oscillator to the atomic center of mass mode. In order to employ the atoms as a coolant, we apply laser cooling techniques to the trapped atomic ensemble.

Such laser cooling for trapped atoms was already studied in several experiments. In [97], atoms were trapped in a 1D optical lattice, and more than every second atom was cooled to the vibrational ground state with superimposed polarization gradient cooling. In a later experiment, the same achieve substantially unity occupation in a 3D optical lattice [98] by employing an optimzed experimental sequence. Applying raman-sideband cooling to atoms trapped in a 3D optical lattice allows one [99] to cool 3×10^8 atoms to a phase space density of $n\lambda_{dB}^3 = 1/500$. This is three orders of magnitude beyond the density that can be achieved with MOT or Molasses cooling techniques in free space, which is due to the reduction of light assisted collisions, which is one of the dominant loss mechanisms in such experiments [100].

However, the trap lifetimes in our experiment are rather short in comparison to the decay time of the amplitude of our mechanical oscillator, because we trap atoms in a 1D optical lattice instead of a 3D optical lattice as in [97, 98, 99]. Hence, it is beneficial to apply a cooling scheme which allows to replenish the lattice permanently with atoms. We achieve this in our experimental situation where the optical access is limited, by superimposing a MOT to the 1D optical lattice. This allows us to trap continuosly a high number of atoms during the experiment despite of the short trap lifetime.

3 Mechanical coupling via the surface potential

In this chapter, I report on the experiment where we couple the motion of a micromechanical cantilever oscillator to the motion of ultracold atoms. An atom chip is used to prepare and position a Bose-Einstein condensate (BEC) close to the cantilever which is glued onto an atom chip. The coupling mechanism between the oscillator and the BEC relies on surface forces such that functionalization of the mechanical oscillator with a mirror or a magnet is not required. Since functionalization of small mechanical oscillators is technically challenging, this coupling mechanism could be suited for coupling ultracold atoms to molecular scale oscillators, which could be interesting because of their small effective mass.

The content of this chapter was published in [81]. In the following, I describe briefly the theoretical foundations, the setup and the central results. For more detailed information, the reader is referred to the thesis of David Hunger [54] which gives a thorough description of the subject.

3.1 Coupling scheme

We use an atom chip to prepare and trap a BEC in a magnetic trapping potential close to a micromechanical cantilever. The trapping potential is generated by a current flow through microfabricated wires on an atom chip, and a superimposed homogeneous external field. The experimental situation is shown in Fig. 3.1. The tip of the mechanical oscillator is positioned at a distance z_c above the wire structure which is employed to provide the magnetic trapping potential for a BEC at a distance d from the equilibrium position of the mechanical oscillator. An atom in the vicinity of the oscillator surface experiences an attractive surface potential. For distances d in the micrometer range, the strong potential gradient of the surface potential modifies the magnetic trap considerably. Motion of the mechanical oscillator thus leads to a modulation of the trapping potential and couples to the motion of trapped atoms.

In the following, the potential between a neutral atom and the surface is described. Excitation modes of a BEC due to the modulation of the trapping potential are

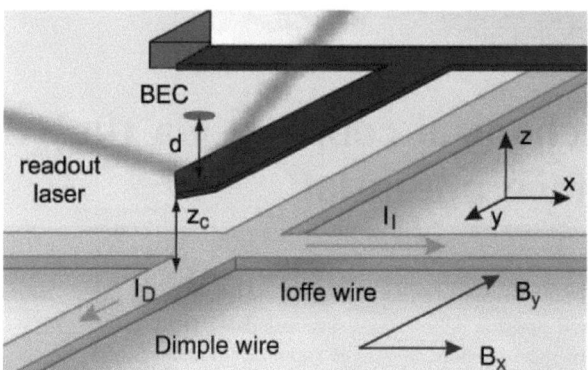

Figure 3.1: A BEC is trapped close to a mechanical oscillator which is glued onto an atom chip. The surface potential modifies the magnetic trapping potential and thus couples the mechanical oscillator to the BEC. A readout laser beam is used to monitor the motion of the mechanical oscillator independently.

briefly reviewed.

3.1.1 Surface potentials

The attractive surface potential which acts onto neutral atoms close to a surface can have many origins. We focus our description onto the Casimir-Polder potential and the adsorbate potential, which arises from a spatially inhomogeneous distribution of atoms adsorbed on a surface. Other contributions could be magnetic impurities in the mechanical oscillator, or inhomogeneous electric potentials arising from static electric charges on a dielectric surface.

Casmir-Polder potential The interaction of an atom and a dielectric surface can be regarded as the interaction of the fluctuating dipole moment of an atom with fluctuating dipole moments on the surface. The resulting van-der-Waals potential is given by $U_{vdW}(z) = -C_3/z^3$. For distances $z \gg \lambda/2\pi$, retardation has to be taken into account. In the retarded regime, Casimir and Polder derived the interaction potential between an atom and a perfectly conducting half-space and found [101] the total interaction energy

$$U_{CP}(z) = -\frac{3\hbar c \alpha_0}{8\pi z^4}, \tag{3.1}$$

3.1 Coupling scheme

with the static atomic polarizability α_0. The theoretical description is adapted for a dielectric half-space and an atom [102]

$$U_{CP}(z) = -\frac{3\hbar c \alpha_0}{8\pi z^4} \frac{\epsilon_r - 1}{\epsilon_r + 1} \phi(\epsilon_r) = -\frac{C_4}{z^4} \quad (3.2)$$

with $\phi(\epsilon_{r,SiN}) = 0.77$ for $\epsilon_{r,SiN} = 4.08$ [103].

The scaling of the surface potential of a dielectric surface changes if one considers not a half-space, but a thin dielectric layer. In the limit where the distance z is large compared to the thickness of the dielectric layer, the van-der-Waals potential falls off with z^{-4} and the Casimir-Polder potential scales with z^{-5}. In contrast, metallic walls with finite thickness can be described with the result obtained for a metallic half-space, as long as the conductivity of the metal allows for 'metallic' boundary conditions [104, 105, 106].

Adsorbate potential In contrast to the van-der-Waals and Casimir-Polder potential which arise always close to surfaces, adsorbate potentials rely on a spatially inhomogeneous distribution of atoms adsorbed on the surface. In our experiment, the mechanical oscillator is in a ultra high vacuum chamber with an enhanced background pressure of Rb atoms which can stick to surfaces. Moreover, in our experimental situation Rb atoms lost from the magnetic trap due to the influence of the attractive surface potential are accelerated towards the surface. An adsorbed Rb atom partially transfers the valence electron to the surface, and forms a permanent electric dipole moment [107]. The stray field resulting from adsorbed atoms polarizes Rb atoms trapped nearby, and thus contributes to the effective surface potential. In particular, an inhomogeneous distribution can have a significant effect onto the trapping potential and both scaling and strength depend strongly on the distribution of adsorbed atoms. Estimates [54] assume that the inhomogeneous distribution of adsorbed atoms follows the shape of the BEC in a trap close to the metallic surface, and show that the impact of the adsorbate potential onto the trapping poteential can exceed the impact of the Casimir-Polder potential at micrometer distances, which is the relevant length scale in our experiments.

3.1.2 Effect of the surface potential onto trapped atoms

As discussed above, the trapping potential is the sum of the magnetic potential U_m and the surface potential U_s which includes contributions from the Casimir-Polder potential U_{CP} and an additional potential U_{ad} due to e.g. adsorbed Rb atoms on or contamination of the mechanical oscillator. U_{grav} accounts for gravitation which is directed perpendicular to the surface. The combined potential reads

Mechanical coupling via the surface potential

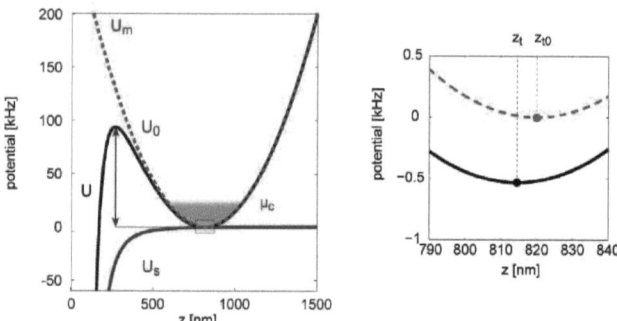

Figure 3.2: The magnetic trapping potential U_m is deformed by the surface potential U_s, resulting in the effective potential U with trap frequency $\omega_z/2\pi = 10$ kHz. The shaded area indicates the extension of a BEC of $N = 600$ atoms. The right hand side picture illustrates the minimum shift. Picture adapted from [54].

$$U[z] = U_m + U_{CP} + U_{ad} + U_{grav} \qquad (3.3)$$
$$= \frac{1}{2}m\omega_{z,0}^2(z - z_{t,0})^2 - \frac{C_4}{(z-z_c)^4} + U_{ad}[z - z_c] + mgz, \qquad (3.4)$$

with the surface position z_c and the magnetic trap minimum $z_{t,0}$. The modification of the harmonic trapping potential due to the surface potential $U_s = U_{CP} + U_{ad}$ is illustrated in Fig. 3.2. The depth of the effective trapping potential is reduced by the attractive surface potential, when the trap minimum is in the vicinity of the surface. If the energy of a trapped atom exceeds the remaining trap depth, the atom is 'suddenly' lost from the trapped ensemble within one oscillation cycle.

Excitation of the mechanical oscillator to an amplitude a results in a modulation of the trap frequency and a shift of the trap minimum. These two effects are employed to couple the motion of the mechanical oscillator to the motion of a trapped BEC in the experiments described in section 3.3.

- **Modulation of the trap frequency:** The curvature of the surface potential depends strongly on the distance to the surface. Modulation of the distance between the mechanical oscillator and the trapped ensemble results in a variation of the curvature which translates into a change of the trap frequency [102]

$$\omega_z^2 = \omega_{z,0}^2 + \frac{1}{m}\frac{\partial^2 U_s}{\partial z^2}. \qquad (3.5)$$

3.1 Coupling scheme

- **Modulation of the trap minimum:** The gradient of the surface potential shifts the minimum $z_{t,0}$ of the magnetic trap to a new minimum position z_t when the surface is approached

$$z_t \approx z_{t,0} - \frac{1}{m\omega_z^2}\frac{\partial U_s}{\partial z}, \tag{3.6}$$

as illustrated in Fig. 3.2 (right).

3.1.3 Mechanical modes of Bose-Einstein condensates

The motion of the mechanical oscillator couples to excitation modes of the trapped BEC via modulation of the minimum position and the trap frequency. This excites the center of mass mode, the radial breathing mode, the quadrupole mode and gives rise to an energy transfer into the respective mode.

BEC in the Thomas-Fermi regime Bose-Einstein condensation sets in [49, 108, 109], when the thermal de-Broglie wave length λ_{dB} is of the same order as the interatomic distance, which is fulfilled if the condition $n\lambda_{dB}^3 \simeq 2.612$ with the atomic peak density n is met.

The BEC is described in mean-field theory by the Gross-Pitaevskii equation (GPE)

$$\left(-\frac{\hbar^2}{2m}\nabla^2 + U_{ext}(\boldsymbol{r}) + gN_0|\phi(\boldsymbol{r})|^2\right)\phi(\boldsymbol{r}) = \mu_c \phi(\boldsymbol{r}), \tag{3.7}$$

with $g = 4\pi\hbar^2 a_s/m$ and the number of atoms N_0. The harmonic oscillator length is given by $a_{ho} = \sqrt{\hbar/m\omega_{ho}}$, with the geometric average of the oscillation frequencies given by $\omega_{ho} = (\omega_x \omega_y \omega_z)^{1/3}$, with indices $i = x,y,z$ denoting the trap axes. The interaction of the atoms is modelled with a contact potential with the s-wave scattering length a_s

$$U_{int}(\boldsymbol{r}_i - \boldsymbol{r}_j) = \frac{4\pi\hbar^2 a_s}{m}\delta(\boldsymbol{r}_i - \boldsymbol{r}_j). \tag{3.8}$$

In the limit $N_0 a_s/a_{ho} \gg 1$, the description of the BEC can be simplified in the Thomas-Fermi (TF) approximation under neglection of the kinetic energy term of the GPE in equation 3.7. The density profile reflects the shape of the trapping potential

$$n_c(\boldsymbol{r}) = |\phi(\boldsymbol{r})|^2 = \max\{0, (\mu_c - U_{ext}(\boldsymbol{r}))/g\}. \tag{3.9}$$

The chemical potential μ_c and the TF radii of the condensate are given by

$$\mu_c = \frac{\hbar\omega_{ho}}{2}\left(\frac{15a_s N_0}{a_{ho}}\right)^{2/5} \tag{3.10}$$

$$R_{TF,i} = \sqrt{\frac{2\mu_c}{m\omega_i^2}}. \tag{3.11}$$

In order to describe the situation in our experiment with small atom numbers $N_0 \simeq 10^3$, the TF approximation can be extended [110].

Excitation of a BEC The frequencies of excitation modes of a non-interacting BEC are given [111, 112] by $\omega(n,l) = \omega_{ho}(2n+l)$, where n and l are the principal and the angular momentum quantum number. Excitations of a BEC of interacting atoms are treated within the Bogoliubov theory where the interaction is modelled in a way which is similar to the description of a superfluid in the hydrodynamic limit. The frequencies of the lowest collective excitation modes of a BEC in a cigar shaped trapping potential with $\omega_\perp \gg \omega_x$ above the dipole mode at $\omega = \omega_\perp$ are given by

$$\omega_{l=2,m=0} = 2\omega_\perp, \tag{3.12}$$
$$\omega_{l=2,m=0} = \sqrt{5/2}\omega_x, \tag{3.13}$$

where the axial component of the angular momentum is described with the quantum number m. The radial compression mode (equation 3.12) corresponds to the parametric resonance of a thermal non-interacting ideal gas. The frequency of the quadrupole mode ($l=2, m=2$) (3.13) can be calculated [111, 112, 110] from the kinetic and potential energies

$$\omega_{l=2,m=2} = \sqrt{2}\omega_\perp\sqrt{1 + E_{kin,\perp}/E_{pot,\perp}}. \tag{3.14}$$

In section 3.3.3, the motion of the mechanical oscillator is transduced via the surface potential and modulates the trap such that the center of mass mode, the radial compression mode and the quadrupole mode can be resolved in a spectroscopy of the trapped BEC. The excitation of all modes of a trapped BEC dephases quickly due to a strong anharmonic deformation of the magnetic trapping potential by the superimposed surface potential.

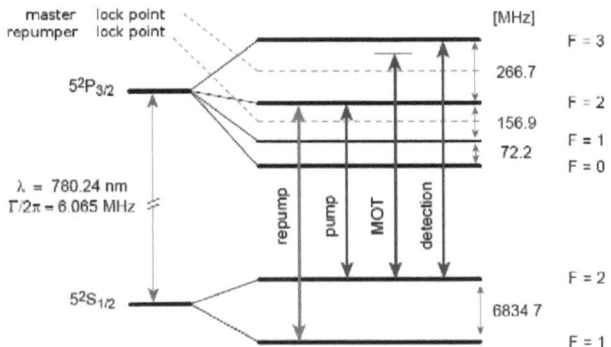

Figure 3.3: Relevant wavelengths for adressing ^{87}Rb. Figure adapted from [54].

3.2 Atom chip setup

In this thesis an atom chip setup similar to the setups described in [113, 91] was built up. The atom chip was designed for the experiments which are described in this chapter. We further use the setup as a source of ultracold atoms in the experiments described in chapter 4. This specific setup is already described in detail in the thesis of David Hunger [54]. I give a brief description of the central components of the experimental setup.

3.2.1 'Standard' setup

Laser system The laser system provides the light which is used for the mirror-MOT cooling beams, pumping, repumping and detection beams. In our laser system, we employ diode lasers which are locked to an optical transition with doppler-free saturation spectroscopy and provide frequency stabilized laser light at the relevant wave lengths close to the D_2 line of ^{87}Rb as shown in Fig. 3.3. The required frequencies are derived with (double pass) AOMs from the lasers which are frequency stabilized to the lockpoints. The light is coupled to polarization maintaining single-mode optical fibers for spatial mode cleaning, collimated and sent into the vacuum chamber.

Vacuum system The vacuum system is assembled with standard CF components (see Fig.3.4), and after initial pump down with a turbo pump and baking, an ion pump maintains a pressure of $p \simeq 5 \times 10^{-10}$ mbar, assisted by Ti sublimation once a year. The speciality of the atom chip setups in our group is the pyrex cell attached to

38 Mechanical coupling via the surface potential

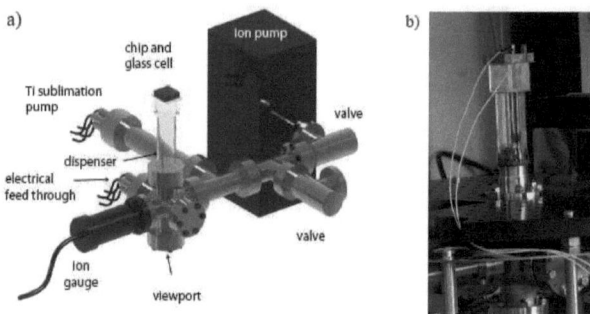

Figure 3.4: (a) Schematic of the the vacuum system. (b) Pyrex glass cell. Figure taken from [54].

the steel vacuum chamber with a glass-to-metal transition. The pyrex cell is closed with an atom chip which is glued onto the top side with an epoxy glue[1]. Apart from good optical access, this construction provides vacuum feedthroughs via chip wires crossing the adhesive area. Rubidium vapour is provided from dispensers based on Rb chromate[2] which are operated close to the threshold.

Magnetic fields The magnetic field landscape is provided by the current flow through microfabricated wires, and the superimposed homogeneous magnetic fields. In addition to the microfabricated atom chip wires and the water cooled (anti) Helmholtz coils surrounding the glass cell, we employ a water cooled copper U-bar which is mounted on the backside of the atom chip in order to provide a quadrupole type magnetic field for the first mirror-MOT stage. The prerequisite for stable magnetic fields are stable current sources. Several thereof were developed in our group and combine high stability, low noise and fast switching times [114].

3.2.2 Atom chip and BEC production

The atom chip used for the experiments consists of a base and an experiment chip, which is glued on top of the base chip with an epoxy glue[3]. The assembly is shown in Fig. 3.5. The gold wire structures on the chip surfaces are fabricated with optical lithography and electroplating technique. Details of this process can be found in [113]. The wires on the experiment chip are electrically connected to the wires on the base chip with bond wire connections. The mirror used for the mirror-MOT is

---
[1] Epo-Tek 353 ND
[2] SAES Rb/NF/3.4/12 FT10+10
[3] Epo-Tek H77S

3.2 Atom chip setup

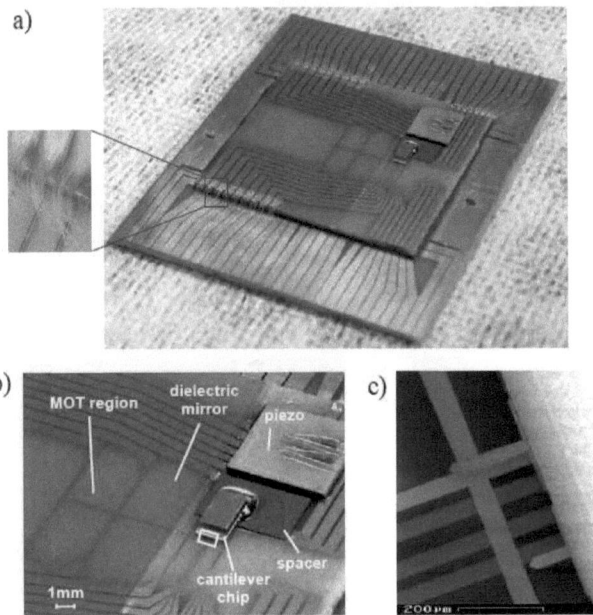

Figure 3.5: (a) Photograph of the atom chip used in the experiment. (b) An AFM chip supporting mechanical oscillators is glued onto the atom chip. (c) SEM micrograph of the mechanical oscillators sticking out of the AFM chip.

a dielectric mirror glued on top of the experiment chip.

The mechanical oscillators used in this experiment are commercially available[4] mechanical oscillators used for atomic force microscopy (AFM). The mechanical oscillators, 'cantilevers', stick out from the edge of a carrier chip. We glue the carrier chip onto the experiment chip, sandwiching a spacer chip of 47 µm thickness in between such that the mechanical oscillators are positioned $\simeq$ 60 µm above the microfabricated wires on the experiment chip. The amplitude a of the mechanical oscillator can be excited by driving a nearby piezo with a signal generator The piezo is electrically contacted with conducting glue[5] from the back and with bond wires from the top.

---

[4] NanoAndMore GmbH, 35578 Wetzlar, Germany
[5] Epo-Tek H20E

The experimental sequence starts with a mirror-MOT [86] at several mm distance from the chip surface using the copper U-bar carrying a current of 55 A, and an external field of $\boldsymbol{B} = (0.0, 9.7, 3.0)$ Gauss. ^{87}Rb atoms are collected from the background gas and transferred to a series of overlapping mirror-MOTs which transfer the atom cloud close to the chip surface. The role of the copper U-bar is taken over by U-shaped wires on the base chip. To achieve a high density, the atom cloud is compressed in a MOT with increased detuning prior to the molasses cooling phase. It is notable, that only the first MOT runs in steady state, while the subsequent MOT stages are operated in a transient regime which is optimized to transport a high number of ultracold atoms with a high density close to the chip surface. The temperature of the ensemble is reduced to $\simeq 10$ µK in the molasses phase of 2.8 ms and a detuning of $\Delta = -11\Gamma_{se}$, with the natural linewidth $\Gamma_{se} = 2\pi \times 6$ MHz.

The molasses cooled atoms are optically pumped to $|F = 2, m_F = 2\rangle$, and loaded into a Ioffe-Pritchard trap which is provided by means of a Z-shaped wire structure on the experiment chip. The transport from the mirror-MOT region to the mechanical oscillator over a distance of 6.4 mm employs a wire based waveguide with a superimposed quadrupole field as discussed in 2.2.1. After the transport, the atoms are loaded into a Ioffe-Pritchard type trapping potential, which is modulated by currents perpendicular to the central wire. In this dimple trap, the atoms are Bose-Einstein condensed in three stages of evaporative cooling, and pure BECs of $N \simeq 2000$ atoms are produced at a distance of $d = 16.6$ µm from the mechanical oscillator. The density distribution of the atomic cloud is detected with (destructive) absorption imaging [49].

3.3 Measurements

We characterize the surface potential of the mechanical oscillator with a BEC in order to calibrate the distance between the oscillator and the atomic ensemble. In subsequent experiments, we position the BEC in the range where the surface potential modifies the magnetic trapping potential significantly, and use this to image the fundamental mode resonance with the BEC. We further use the modulation of the trapping potential to perform a spectroscopy of the trapped BEC. We demonstrate controlled excitation of the center of mass mode, the radial breathing mode and the quadrupole mode of the BEC.

3.3.1 Characterization of the surface potential

The position of the magnetic trap minimum is simulated from the current flow through wires on the atom chip and from the external, magnetic fields [54]. In order

3.3 Measurements

to locate the mechanical oscillator we ramp the magnetic trap minimum adiabatically towards the mechanical oscillator, hold the BEC at a distance d for a hold time $t_h = 1$ ms and ramp the atoms back into a relaxed trap at a large distance from the surface. These experiments are performed with the non-excited mechanical oscillator[6]. In subsequent shots of the experiment, the distance d is decreased, and we measure the number of remaining atoms in the trap after t_h with absorption imaging. Fig. 3.6 shows the result of such measurements at both sides of the mechanical oscillator. The upper axis gives the distance to the wires on the chip, and we plot the remaining fraction of atoms $\chi = N_r/N$ as a function of the distance d to the chip surface. This can be interpreted as a measurement of the effective height of the mechanical oscillator.

In order to locate the mechanical oscillator, and to obtain the bottom axis with the distance calibration between the mechanical oscillator and the trap minimum we analyze the data as follows. We identify $\chi = 0$ with the position where the trapping potential vanishes. The analysis is based on a model which assumes that all atoms whose energy is larger than the trap depth are suddenly lost from the trapped ensemble when the depth of the potential is reduced below the energy of an atom [115]. The remaining fraction of atoms in this model is $\chi = 1 - \exp(-U_0/k_B T)$. We apply this to model the loss of atoms from the thermal cloud which coexists with the BEC, and we further include losses due to evaporation and tunneling through the potential barrier.

The resulting model is employed to fit to the data in Fig. 3.6. The surface potential on the dielectric backside is assumed to be the Casimir-Polder potential U_{CP} of a thin dielectric slab of silicon nitride (SiN), together with a contribution from the metallic surface on the other side. The effective Casimir-Polder potential is calculated with [116, 104, 105, 106], and found to be 25 % larger than U_{CP} of the dielectric SiN slab alone. For the metallized side, we assume the Casimir-Polder potential of a perfect conductor due to the decent conductivity of the 65nm thick gold-chromium layer. One finds that the data in Fig. 3.6 can not be explained exclusively with the assumption of a Casimir-Potential potential U_{CP}. One has to assume an additional potential $U_{ad} \gg U_{CP}$ on at least one side of the mechanical oscillator. A possible candidate is an attractive potential due to Rb atoms adsorbed on the metallized side of the mechanical oscillator. Estimates [54, 81] show, that one could expect a potential U_{ad} due to inhomogeneously adsorbed atoms with $U_{ad} \gg U_{CP}$ in our experimental situation. With this analysis, one obtains a calibration of the distance d between the mechanical oscillator and the trap minimum. The position of the cantilever is indicated in Fig. 3.6.

---

[6]The thermal amplitude is $a_{th} \approx 0.4$nm, and can not be resolved in measurements with ultracold atoms

Mechanical coupling via the surface potential

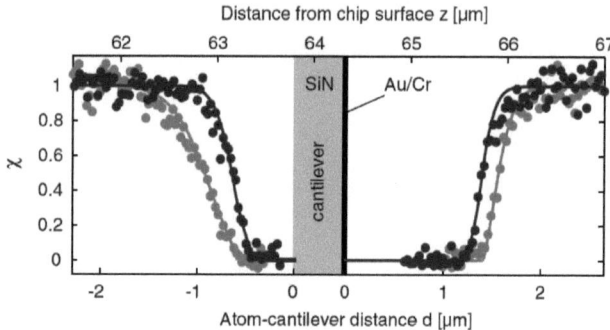

Figure 3.6: Fraction χ of atoms remaining in the trap after a hold time $t_h = 1$ ms of the atomic ensamble at a distance d from the cantilever surface. Dark (light) grey data points correspond to a trap with $\omega_z/2\pi = 10.0$ kHz (5.1 kHz). The lower axis shows the calibration obtained from the analysis (solid lines).

We further determine the positioning reproducibility of a BEC from the atom number noise on the slope at $d = 1.3$ μm, and find an upper limit to the position uncertainty of $\Delta z_{t,0} = 6$nm rms.

3.3.2 Imaging of the mechanical oscillators' resonance

We investigate the coupling of the mechanical oscillator to a trapped BEC positioned at a distance d from the metallized side of the mechanical oscillator. The mechanical oscillator is excited to an amplitude a with the nearby piezo which is driven at the frequency ω_p. The fundamental out-of-plane mode of the mechanical oscillator has an eigenfrequency of $\omega_m \approx 2\pi \times 10$ kHz[7]. The oscillator amplitude is calibrated and permanently monitored with an independent optical readout laser beam at 830nm, which is sufficiently far detuned from the D_1- and D_2-line of ^{87}Rb such that it does not affect the atoms. The laser beam is reflected from the metallized side of the mechanical oscillator such that the angle of the deflection from the rest position is translated into a position shift of the reflected laser beam. This shift is monitored with a position sensitive two quadrant photodetector, and allows one to extract oscillation frequency and amplitude [117].

For studying resonant coupling of the fundamental out-of-plane mode of the me-

---

[7]The eigenfrequency of the fundamental mode decreased over time, which we attribute to aging of the mechanical oscillator.

3.3 Measurements

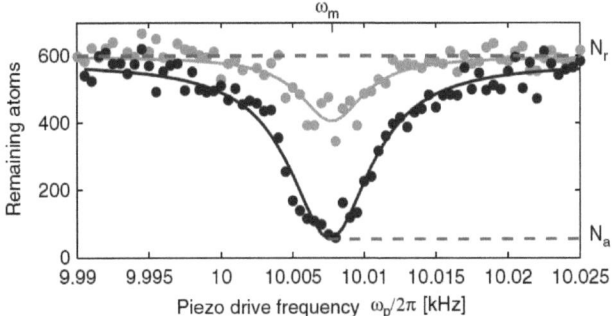

Figure 3.7: Number of remaining atoms after $t_h = 3$ ms in a trap with $\omega_z/2\pi = 10.5$ kHz at a distance $d = 1.5$ μm from the driven mechanical oscillator. The dark (light) grey circles correspond to a cantilever amplitude $a = 120$nm (50nm) on resonance. The solid lines are Lorentzian fits with a 6 Hz FWHM corresponding to the width of the oscillators' resonance.

chanical oscillator to the center of mass mode of the trapped ensemble, the trap is positioned at a distance $d = 1.5$ μm from the equilibrium position of the cantilever. The trap frequency of $\omega_z/2\pi = 10.5$ kHz is chosen such that the resonance condition $\omega_m \approx \omega_z$ is fulfilled. The excitation frequency ω_p is scanned from shot to shot of the experiment, while the magnetic trap is always prepared at the same position. Scanning ω_p effectively results in scanning across the resonance of the mechanical oscillator, yielding a different amplitude in each step of the experiment. The sequence for the detection of the atom number is similar to the experiments described in the previous section, where the atoms are hold for a time t_h at a distance d, and ramped to a relaxed trap for subsequent absorption imaging. Fig. 3.7 shows measured data for two different (peak) amplitudes a of the mechanical oscillator at $\omega_m = \omega_z$.

The enhanced oscillator amplitude on resonance leads to a modulation of the trap minimum position z_t with an amplitude $\delta z_t = 10$nm (4nm) for $a = 120$nm (50nm). This periodic shaking of the trap couples to the center of mass motion of the trapped atomic ensemble. The width of the resonance is in reasonable agreement with the quality factor of the mechanical oscillator which is determined to be $Q = 3200$ in a ringdown measurement of the amplitude. From this alternative measurement, we would expect a FWHM of approximately 10 Hz, while we find a width of 6 Hz from the measurement with the BEC.

This measurement is done for several oscillator amplitudes a. In Fig. 3.8, we plot the contrast $C = (N_r - N_a)/N_r$ and the signal to noise ratio SNR$= (N_r - N_a)/\sigma$, where

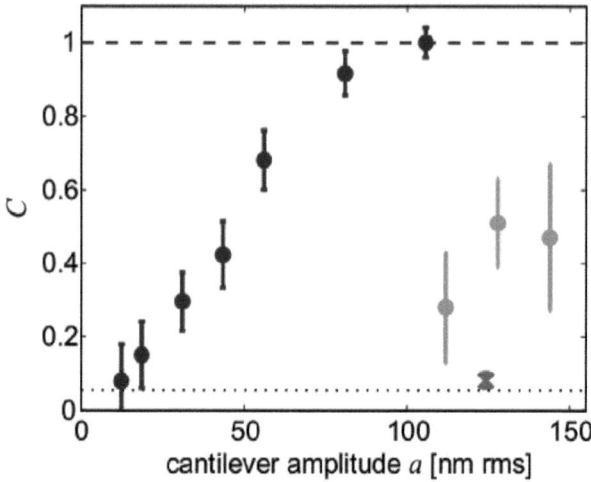

Figure 3.8: Sensitivity of the readout on the metallized side of the mechanical oscillator on resonance with the atomic ensemble (dark grey). The contrast C quantifies the modulation depth of the resonance (see Fig. 3.7). The smallest detectable cantilever amplitude is $a = (13 \pm 4)$nm for SNR= 1 without averaging. Analogue measurements performed on the dielectric side of the oscillator reveal a weaker coupling (light grey). The data point with the smallest error bars was obtained from measurements with averaging in an off-resonant trap with $\omega_z/2\pi =$ 4 kHz. The rms noise of the measurement is indicated with the dotted line.

N_a (N_r) is the number of remaining atoms after a hold time t_h at a distance d for off-resonant (on-resonant) driving. We detect a smallest amplitude $a = (13 \pm 4)$nm *without* averaging and with a SNR= 1. This confirms, that the influence of the thermal motion onto a trapped ensemble is negligible, and that we indeed can assume the mechanical oscillator as a static freestanding structure in section 3.3.1.

We investigate the dependence of the atomic signal on the distance d from the oscillator on the metallized side for a fixed oscillator amplitude $a = 90$nm. Fig. 3.9 shows the contrast C and the signal to noise ratio SNR of the driven (undriven) mechanical oscillator. Intuitively, one would expect the maximum of the SNR at the position of the steepest slope at $d \approx 1.5$ μm in Fig. 3.6. This expectation is confirmed by the measurement shown on the right hand side of Fig. 3.9 in a trap at 10.5 kHz.

3.3 Measurements

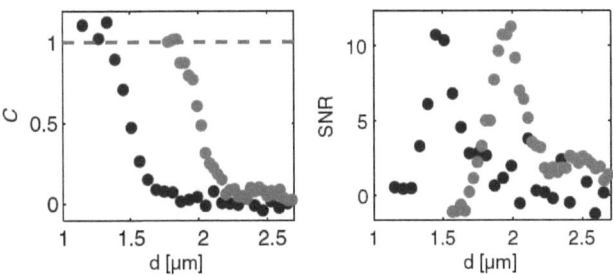

Figure 3.9: Contrast C and signal to noise ratio SNR of the observed atomic signal as a function of d, for constant $a = 90$nm and $\omega_p = \omega_m$. Dark (light) grey data points correspond to $\omega_z/2\pi = 10.5$ kHz (5.0 kHz) and $t_h = 3$ ms (20 ms).

3.3.3 Spectroscopy of the trapped BEC

In the last section, the piezo driving frequency ω_p was scanned from shot to shot of the experiment, allowing to image the resonance of the mechanical oscillator with the atomic ensemble. In this section, the cantilever is always driven on resonance at a fixed amplitude $a = 180$nm. Due to aging of the oscillator, the eigenfrequency of the fundamental mode has decreased over time to $\omega_m/2\pi = 9.68$ kHz. The BEC is prepared at the metallized side of the oscillator, and the trap frequency is varied from shot to shot with the trap position adjusted such that the number of atoms N_r throughover the scan is approximately constant. The measured SNR is shown in Fig. 3.10.

We observe the resonance which corresponds to the excitation of the center of mass mode at $\omega_m = \omega_z$. The modulation amplitude of the trap minimum is $\delta z_t = 7$nm. The peak at $\omega_m = 2\omega_z$ is attributed to the radial breathing mode which is excited by modulation of the trap frequency $\delta \omega_z = 2\pi \times 150$ Hz. The linewidth of the resonance at $\omega_z = \omega_m$ is broadened in contrast to the linewidth of only 60 Hz of the resonance at $\omega_z = 2\omega_m$, which might be due to the larger thermal component in the anharmonic potential with $\omega_z = \omega_m$. On the left hand side of these resonances, we reproducibly observe 'anti' resonances (red arrows in Fig. 3.10), where the SNR is suppressed by a factor of 20. This shows that the coupling can be efficiently controlled by a small detuning of the trap frequency. We also find weaker resonances at $\omega_m = (1.6, 1.8, 2.1, 2.4)\omega_z$ (arrows in Fig. 3.10) and identify the first resonance with the quadrupole mode of the BEC, whose frequency is calculated with equation 3.14. For smaller d we observe a broadening of these resonances, and find that the resonance which corresponds to an excitation of the quadrupole mode at $\omega_m = 1.6\omega_z$ becomes stronger than the resonance at $\omega_m = 2\omega_z$.

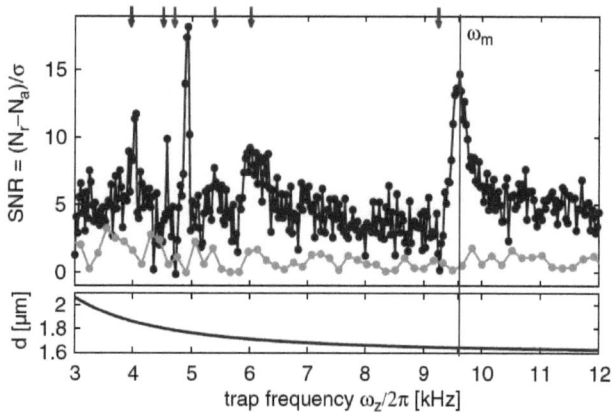

Figure 3.10: Top graph: Response of the atomic ensemble as a function of ω_z for constant $a = 180$nm and $t_h = 20$ ms (dark grey). Light grey: reference measurement without piezo excitation. Bottom graph: Set values of d, chosen such that $N_r(\omega_z) \approx$ const. ($N_r(10\text{ kHz}) = 700$, $N_r(5\text{ kHz}) = 1100$) and that N_a does not saturate.

The experiments presented in this chapter show that it is possible to realize an interaction between a mechanical oscillator and ultracold atoms. However, we did not observe the backaction of the atomic ensemble onto the oscillator. This aspect is investigated in the next chapter, where we observe the backaction onto the mechanical oscillator.

4 Optomechanical coupling via an optical lattice

The focus of the experiment described in this chapter is the observation of backaction of an ensemble of ultracold atoms onto the center of mass mode of a mechanical oscillator. The experiment is facilitated by the long-distance coupling mediated by the optical lattice, which allows one to keep the oscillator and the atoms in individual vacuum chambers. Similar to the experiment in chapter 3, the motion of the oscillator is coupled to the atomic motion which results in a coupling strength $g \propto \sqrt{Nm/M}$. The coupling strength in the experiments in chapter 3 is too small, to resolve the backaction of the BEC onto the mechanical oscillator. The coupling scheme via an optical lattice allows for a sufficient coupling strength to observe the backaction onto a mechanical oscillator due to an increase of the atom number and the mechanical quality factor by 3 orders of magnitude each.

In the following the coupling mechanism via an optical lattice is explained. The experimental implementation of the optical lattice and the readout of the mechanical oscillator is described, and the chapter closes with measurements which demonstrate the observation of backaction. These experiments will be published in [82].

4.1 Coupling scheme

The setup is illustrated in Fig. 4.1. A laser beam comes in from the right hand side and is (partially) retroreflected at a harmonically bound mirror. The reflected beam is overlapped with the incoming beam, and forms a standing wave pattern. The frequency of the laser light is red detuned with respect to an optical transition of ^{87}Rb atoms, which allows to trap atoms in the resulting one dimensional optical lattice potential (1D optical lattice). Motion of the mechanical oscillator shakes the lattice and couples the oscillators' motion to the atomic center of mass (COM) motion. On the other hand, the motion of the atoms in the trapping potential leads to a redistribution of photons between the two running wave components forming the lattice and is thus imprinted onto the power of the laser beam that is retroreflected at the oscillator. The resulting modulation of the radiation pressure constitutes the backaction of the atoms onto the oscillator.

Optomechanical coupling via an optical lattice

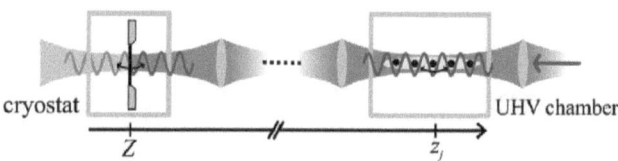

Figure 4.1: A laser beam incoming from the right is partially reflected off a mechanical oscillator and forms a standing wave optical potential for ultracold atoms. The optical lattice provides a long distance coupling, which allows to place the atoms and the mechanical oscillator in different vacuum chambers, or cryostats. Figure taken from [39].

Conventional optical lattice experiments impose high demands onto the mechanical stability of lattice end mirrors, as these experiments require a static potential, whereas we study the situation with an oscillating mirror. We choose a mechanical oscillator as an end mirror which has a reasonable reflectivity such that the reflected beam can be used to form an optical lattice.

In the following, we give a simple picture how the membrane acts onto the atoms and vice versa, treating the light field classically. Radiation pressure induced momentum diffusion processes are as well neglected as retardation effects of the light field. These effects are taken into account by a full quantum theory, which is published in [39] and reviewed in section 4.1.4.

4.1.1 Effect of the mechanical oscillator onto trapped atoms

The situation is simplified in assuming that the coupled system can be described in the picture of two harmonic oscillators, see Fig. 4.2. A displacement δz_m of the mirror leads to a displacement of the potential minima of the 1D optical lattice. The displacement results in a change of the potential for a trapped atom, and thus influences the further motion of the atom. This constitutes a coupling of the mirror motion to the motion of atoms which are trapped in the 1D optical lattice. The equations of motion for an atom are derived in order to obtain an expression for the coupling strength of the mechanical oscillators' motion to the motion of a trapped atom.

Starting from a harmonic trapping potential, we write down the Hamiltonian for the j-th atom with trap frequency ω_{at}

4.1 Coupling scheme

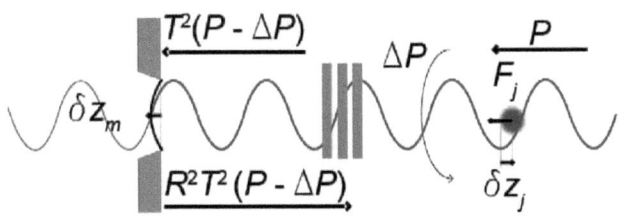

Figure 4.2: Optomechanical coupling of atomic ensemble and membrane oscillator.

$$H_j = \frac{p_j^2}{2m} + \frac{1}{2}m\omega_{at}^2(\delta z_j - \delta z_m)^2, \tag{4.1}$$

where δz_j (δz_m) are displacements of the atoms (membrane) from the equilibrium position. The second term which describes the potential energy, accounts for the influence of the displacement δz_m of the mechanical oscillator onto the potential which the j-th atom experiences. The equations of motion are given by

$$\dot{p}_j = -\frac{\partial H_j}{\partial(\delta z_j)} = -m\omega_{at}^2 \delta z_j + m\omega_{at}^2 \delta z_m \tag{4.2}$$

$$\delta \dot{z}_j = \frac{\partial H_j}{\partial p_j} = \frac{p_j}{m}. \tag{4.3}$$

The force $F_{m \to j}$ onto an atom due to a displacement of the mirror by δz_m, can be extracted from the first equation

$$F_{m \to j} = m\omega_{at}^2 \delta z_m. \tag{4.4}$$

The equations of motion for all atoms N of the ensemble are summed

$$\sum_j \dot{p}_j = -m\omega_{at}^2 \sum_j \delta z_j + Nm\omega_{at}^2 \delta z_m \tag{4.5}$$

$$\sum_j \delta \dot{z}_j = \frac{1}{m}\sum_j p_j, \tag{4.6}$$

and quantized with dimensionless variables $x_{at} = (a_{at}^\dagger + a_{at})$ and $p_{at} = i(a_{at}^\dagger - a_{at})$ which describe the COM motion of the trapped atomic ensemble with $a_{at} = \sum_j a_j/\sqrt{N}$ and $\left[a_{at}, a_{at}^\dagger\right] = 1$. The mirror position and momentum are defined analogously by $x_m = (a_m^\dagger + a_m)$ and $p_m = i(a_m^\dagger - a_m)$.

With $\sum_j \delta z_j = l_{at} \sum_j (a_j^\dagger + a_j)$, one finds

$$\sum_j \delta z_j = \sqrt{N} l_{at} x_{at} \tag{4.7}$$

$$\sum_j p_j = \sqrt{N} m \omega_{at} l_{at} p_{at} \tag{4.8}$$

$$\delta z_m = l_m x_m \tag{4.9}$$

$$q_m = M \omega_m l_m p_m \tag{4.10}$$

where $l_{at} = \sqrt{\hbar/2m\omega_{at}}$ and $l_m = \sqrt{\hbar/2M\omega_m}$ are the ground state spread of the trapped atom and the harmonically bound mirror, respectively, and q_m is the membrane momentum.

We rewrite the equations of motion

$$\dot{p}_{at} = -\omega_{at} x_{at} + 2g_{m \to at} x_m \tag{4.11}$$

$$\dot{x}_{at} = \omega_{at} p_{at}, \tag{4.12}$$

and with the assumption that the two systems are nearly resonant, i.e. $\omega_{at} \approx \omega_m$, one obtains the coupling constant on resonance $g_{m \to at}$

$$g_{m \to at} = \frac{\omega_{at}}{2} \sqrt{\frac{Nm}{M}}. \tag{4.13}$$

From this equation one infers, that the coupling strength depends on the factor $\sqrt{Nm/M}$. In the situation of chapter 3, and in this chapter the mass of the mechanical oscillator exceeds the mass of the atomic ensemble by several orders of magnitude, and this imposes a stringent limit onto the coupling strength. If the resonance condition $\omega_{at} = \omega_m$ is fulfilled an oscillation of the mirror leads to a resonant energy transfer to the atomic ensemble, and to the excitation of the atomic COM mode.

4.1.2 Effect of the atoms onto the mechanical oscillator

In the last section, we have derived an expression for the coupling strength between the motion of a mechanical oscillator and an atomic ensemble. Now, we adress the question how the motion of the mechanical oscillator is influenced by the atomic ensemble. We consider the situation illustrated in Fig. 4.2. An incoming light beam from the right hand side travels to the left with the power P and wave vector $-k$, with $\omega = ck$. The beam is sent from the optical lattice position to the membrane through an optical system which consists of mirrors, lenses, polarization optics and vacuum windows, which has an amplitude transmittivity $|T| < 1$ on the path between Atoms and membrane. The beam is partially reflected at the membrane, which has an amplitude reflectivity R. The 1D optical lattice potential is formed by the interference of the incoming beam and the reflected beam. If there are no atoms in the optical lattice, the beam which is reflected at the membrane has a beam power of $|R|^2|T|^2P$ right after the membrane. In the following, we study how the presence of atoms which are trapped in the lattice changes the situation. A harmonically bound atom oscillates in the trapping potential V and experiences a restoring force

$$F_j = \dot{p}_j = -\frac{\partial V}{\partial (\delta z_j)} = -\omega_{at}^2 m_{at} z_j \qquad (4.14)$$

The force increases linearly with the distance to the minimum of the trapping potential, and is directed towards this minimum. The physical origin of this restoring force is the momentum transfer associated with the redistribution of photons between the two running wave components of the 1D optical lattice [118]. The j-th atom redistributes photons at a rate $\dot{n}_j$ in the beam with wave vector $+k$. With the atomic momentum change of $-2\hbar k$ per redistribution event, the restoring force F_j is given by

$$F_j = -2\hbar k \dot{n}_j \qquad (4.15)$$

For N trapped atoms we calculate a total photon exchange rate with

$$\dot{n} = \sum_j \dot{n}_j = -\frac{1}{2\hbar k}\sum_j F_j \qquad (4.16)$$

The power of the beam with wave vector $+k$ which is reflected at the membrane is changes by

$$\Delta P = \hbar\omega \dot{n} = -\hbar\omega \frac{1}{2\hbar k}\sum_j F_j = -\frac{c}{2}\sum_j F_j, \qquad (4.17)$$

when it passes the optical lattice, and the power of the incoming beam is decreased by $-\Delta P$. The value of ΔP depends on the restoring forces acting onto an atomic ensemble which is displaced from the minimum of the trapping potential. In the situation considered above, the power P of the incoming beam changes to $P - \Delta P$, when the beam passes the optical lattice position, and is further decreased to $T^2(P-\Delta P)$ due optical losses between the atoms and the membrane. The reflected beam has a power of $P_R = |R|^2|T|^2(P-\Delta P)$ right after the membrane.

If the trapped ensemble performs a COM motion in the potential, the beam powers are periodically modulated. Experiments show that the power modulation can be significant [119, 120]. According to [119], the power modulation of a lattice laser beam due to the center of mass motion of a trapped ensemble can be estimated by $|\Delta P/P| \approx N(\delta z)\Gamma_{se}10^{-5}/\lambda\Delta$ for a large detuning Δ and a small displacement δz. For experimental parameters similar to the ones used in section 4.3 with $N = 1 \times 10^6$ atoms, a displacement of $\delta z = 0.5$nm, and the natural linewidth $\Gamma_{se} = 2\pi \times 6$ MHz of the D_2 line of ^{87}Rb, a lattice laser wavelength $\lambda = 780$nm and a detuning of $\Delta = -2\pi \times 20$ GHz, one expects a relative power modulation $\Delta P/P = 1 \times 10^{-5}$.

The incoming beam exerts a static radiation pressure onto the mechanical oscillator with a modulation on top due to the component $|T|^2|R|^2\Delta P$. Note, that ΔP is fixed for a given displacement of the atoms in the trapping potential wit fixed ω_{at}. The backaction of the atoms onto the membrane is reduced by a factor $|T|^2|R|^2$, which accounts for photons which do not contribute to the radiation pressure onto the membrane, as they are lost in the optical path or transmitted through the membrane. Each of the photons which is reflected at the membrane transfers a momentum of $-2\hbar k$ to the mechanical oscillator. Hence, the total radiation pressure acting onto the membrane is

$$F_m = \frac{P_R}{\hbar\omega}(-2\hbar k) = -\frac{2}{c}|T|^2|R|^2(P-\Delta P), \qquad (4.18)$$

and with 4.17 the modulation of the radiation pressure force is given by

$$\Delta F_m = \frac{2}{c}|T|^2|R|^2\Delta P = -|T|^2|R|^2\sum_j F_j. \qquad (4.19)$$

We assume a harmonic trapping potential for the atoms with $F_j = -m\omega_{at}^2\delta z_j$ so that

$$\Delta F_m = |T|^2|R|^2 m\omega_{at}^2 \sum_j \delta z_j \qquad (4.20)$$

4.1 Coupling scheme

This expression for ΔF_m allows one to write down the equations of motion for the membrane

$$\dot{q}_m = -M\omega_m^2 \delta z_m + \Delta F_m \tag{4.21}$$
$$\delta \dot{z}_m = \frac{q_m}{M}, \tag{4.22}$$

and to sum and quantize as in the previous section

$$\dot{p}_m = -\omega_m x_m + 2g_{at \to m} x_{at} \tag{4.23}$$
$$\dot{x}_m = \omega_m p_m \tag{4.24}$$

In the near resonant case with $\omega_{at} \approx \omega_m$ one finds the coupling constant $g_{at \to m}$

$$g_{at \to m} = |T|^2 |R|^2 \frac{\omega_{at}}{2} \sqrt{\frac{Nm}{M}}. \tag{4.25}$$

Comparison to the coupling constant which is derived in the previous section reveals an asymmetric coupling

$$g_{at \to m} = |T|^2 |R|^2 g_{m \to at}. \tag{4.26}$$

In a classical picture of the coupling mechanism, the mechanical oscillator and the atomic ensemble can be each modelled as balls which are moving in bowls with harmonic curvature due to gravitation. The coupling would be mediated by an inflexible rod which is rigidly connected to each of the bowls. The result above shows that this analogy does not hold. Although the light field acts like a transfer rod in the sense that it provides coupling, the asymmetry of the coupling constants due to the finite reflectivity can not be understood in this classical analogon.

4.1.3 Backaction of the atoms onto the damping of the membrane

In the two previous sections we found that the action of the membrane onto the atoms and vice versa can be described by asymmetrically coupled harmonic oscillators. In this section, we derive an explicit expression which quantifies the backaction of the atoms onto the damping of the membrane. We start from the equations of motion 4.5, 4.6, 4.21, 4.22, and add damping terms $-\gamma_{at} \sum_j p_j$ $(-\gamma_m q_m)$ to the right hand side of equations 4.5 (4.21).

The equations of motion are inserted into the time derivatives $\dot{a}_{at}$ and $\dot{a}_m$, which leads to the equations of motion

$$\dot{a}_{at} = -i\omega_{at}a_{at} + \frac{\gamma_{at}}{2}(a_{at}^+ - a_{at}) + ig(a_m^+ + a_m) \qquad (4.27)$$

$$\dot{a}_m = -i\omega_m a_m + \frac{\gamma_m}{2}(a_m^+ - a_m) + ig|T|^2|R|^2(a_{at}^+ + a_{at}) \qquad (4.28)$$

with the coupling constant $g = \frac{\omega_{at}}{2}\sqrt{\frac{N_m}{M}}$ near resonance $\omega_{at} \approx \omega_m$. In order to simplify the problem within the rotating wave approximation (RWA), the quadratures a_{at} and a_m are transformed in a frame which is co-rotating with the membrane at frequency ω_m

$$a_{at} = ce^{-i\omega_m t} \qquad (4.29)$$
$$a_m = de^{-i\omega_m t} \qquad (4.30)$$

Inserting this into equations 4.27, 4.28 and defining a detuning $\delta = \omega_{at} - \omega_m$ yields

$$\dot{c} = \frac{\gamma_{at}}{2}(c^+ e^{2i\omega_m t} - c) + ig(d^+ e^{2i\omega_m t} + d) - i\delta c \qquad (4.31)$$

$$\dot{d} = \frac{\gamma_m}{2}(d^+ e^{2i\omega_m t} - d) + ig|T|^2|R|^2(c^+ e^{2i\omega_m t} + c) \qquad (4.32)$$

and with the assumption $\omega_m \gg \delta, g, \gamma_{at}, \gamma_m$ (=RWA)

$$\dot{c} \approx -\frac{\gamma_{at}}{2}c + igd - i\delta c \qquad (4.33)$$

$$\dot{d} \approx -\frac{\gamma_m}{2}d + ig|T|^2|R|^2 c. \qquad (4.34)$$

The damping of the atomic ensemble in our experimental situation is dominated by dephasing of the atomic COM motion due to the spread of atomic vibration frequencies due to the gaussian lattice laser profile and the motion in the anharmonic potential. From the assumption $\gamma_{at} \gg \gamma_m, g$, we found that $\dot{c} = 0$, and equation 4.33 simplifies to

$$0 = -\frac{\gamma_{at}}{2}c + igd - i\delta c \qquad (4.35)$$

which can be transformed

4.1 Coupling scheme

$$c = \frac{ig}{i\delta + \gamma_{at}/2}d \qquad (4.36)$$

and inserted into 4.34

$$\dot{d} = -\left(\frac{\gamma_m}{2} + \frac{\gamma_{at}}{2}\frac{g^2|T|^2|R|^2}{(\gamma_{at}/2)^2 + \delta^2}\right)d + i\delta\frac{g^2|T|^2|R|^2}{(\gamma_{at}/2)^2 + \delta^2}d, \qquad (4.37)$$

which is solved by

$$d(t) = d_0 e^{-\frac{\Gamma_{pop}}{2}t}e^{i\Omega t} \qquad (4.38)$$

with

$$\Gamma_{pop} = \gamma_m + \gamma_{at}\frac{g^2|T|^2|R|^2}{(\gamma_{at}/2)^2 + \delta^2}, \qquad (4.39)$$

$$\Omega = \delta\frac{g^2|T|^2|R|^2}{(\gamma_{at}/2)^2 + \delta^2}. \qquad (4.40)$$

Note, that Γ_{pop} is the damping rate of the membrane population or energy, which is the damping of the amplitude squared. On resonance $\delta = 0$, the damping of the membrane energy is given by

$$\Gamma_{pop} = \gamma_m + \frac{4g^2|T|^2|R|^2}{\gamma_{at}}. \qquad (4.41)$$

The additional damping of the membrane motion due to atoms bound in the optical lattice is given by the second term. This term can be accessed experimentally in ringdown measurements of the membrane amplitude, which decays at a rate

$$\Gamma = \frac{\omega_m}{2Q} + \frac{2g^2|T|^2|R|^2}{\gamma_{at}}. \qquad (4.42)$$

A direct measurement of the second term, $\Delta\gamma = 2g^2|T|^2|R|^2/\gamma_{at}$, is shown in section 4.3.4.

4.1.4 Coupled system – fully quantized theory

The Hamiltonian discussed above included only the energy of the membrane and the atomic ensemble. There, the light field is assumed to provide a potential where the minima are referenced to the membrane position. The light field is treated classically, i.e. quantum fluctuations, which could lead to dissipation are not taken into account. A full quantum treatment of the problem has to include the light field as a dynamical system in the Hamiltonian as well as the various dissipation channels, e.g. shotnoise of the light field. Furthermore, the above analysis does not account for retardation effects due to the long distance coupling. All these effects are included in the full quantum treatment published in [39]. In the following, we briefly sketch the results of this analysis. The treatment does not take transmittivity $|T| < 1$ of the optical system between atoms and membrane into account.

We start from the Master equation, which is derived from the full quantum theory

$$\dot{\rho} = -i[H_{at} + H_m + g x_{at} x_m, \rho] + C\rho + L_m \rho + L_{at}\rho. \tag{4.43}$$

The second term $C\rho$ is responsible for the asymmetry in the coupling strengths, which occurs for $|R| < 1$

$$C\rho = \frac{i(1-R)^2 g}{2} \left([x_m, x_{at}\rho] - [\rho x_{at}, x_m]\right), \tag{4.44}$$

with the amplitude reflectivity R of the mirror. This term is typical for cascaded quantum sytems, where the output of one quantum system forms the input of another quantum system.

The Master equation allows to study the sources of decoherence which arise from dissipation in the mechanical oscillator, the dissipation of the atoms in the lattice, and from the interaction of atoms or mechanical oscillator with the light field. The contributions of the individual sources are described with several Lindblad terms L_m and L_{ui} which are of the general form

$$L_x \rho = \frac{1}{2}\gamma_x D[a]\rho, \tag{4.45}$$

with $D[a]\rho = 2a\rho a^\dagger - a^\dagger a \rho - \rho a^\dagger a$.

- **Dissipation in the mechanical oscillator:** As discussed in the section 2.1.2, the quality factor Q of the mechanical oscillator is limited. Coupling to the thermal bath induces decoherence

4.1 Coupling scheme

$$L_m^{th}\rho = \frac{\gamma_m}{2}(\overline{n}+1)D[a_m]\rho + \frac{\gamma_m}{2}\overline{n}D[a_m^\dagger]\rho, \tag{4.46}$$

which depends on the thermal occupation $\overline{n} = k_B T/\hbar\omega_m$ of the mechanical oscillator which is coupled to a thermal bath with temperature T and the mechanical damping rate of the energy $\gamma_m = \omega_m/Q$.

- **Decoherence and tunable dissipation of the atoms:** The decoherence of the atoms is due to several effects. The spread of vibrational frequencies of atoms in the lattice potential due to the gaussian nature of the lattice beam leads to a dephasing of the center of mass motion as the atoms oscillate at slightly different trap frequencies. This could be reduced by shaping the geometry of the lattice laser beam, e.g. by choosing a flat top beam profile and a large Rayleigh range. Contrarily, the dissipation of the atoms can be controlled by adjusting the cooling rate γ_{at}^{cool} of e.g. Raman side band cooling which is applied to the atomic ensemble. The Lindblad term of this contribution is

$$L_{at}^{cool}\rho = \frac{1}{2}\gamma_{at}^{cool}[a_{at}]\rho. \tag{4.47}$$

Tuning the dissipation allows to change the characteristics of the atomic ensemble in the coupled system, for example to realize sympathetic cooling of the mechanical oscillator for large γ_{at}^{cool}, and to switch to a regime where the coupled system evolves coherently, limited only by decoherence of the mechanical oscillator.

- **Light field induced dissipation of atoms/mechanical oscillator:** Shotnoise of the light field leads to momentum diffusion of mechanical oscillator and atoms. The diffusion rate of the atomic momentum due to spontaneous emission is given by

$$\gamma_{at}^{diff} = (kl_{at})^2 \Gamma_{se} \frac{V_0}{\hbar\delta} \propto \frac{\omega_{rec}}{\omega_{at}} \Gamma_{sc}, \tag{4.48}$$

with the natural linewidth Γ_{se}, the detuning δ from resonance, and the recoil frequency ω_{rec}. γ_{at}^{diff} depends mainly on the parameters of the trapping potential. The momentum diffusion rate of the mechanical oscillator due to the shotnoise in the laser beam with power P is given by

$$\gamma_m^{diff} = \frac{4R^2 P}{Mc^2} \frac{\omega_l}{\omega_m}. \tag{4.49}$$

Ground state cooling In the following, we discuss the parameter regime of a possible experimental realization of this coupling scheme which would allow to reach the quantum ground state of a mechanical mode. A membrane oscillator as described in section 2.1.1 serves as lattice end mirror. In our case, the membrane is an amorphous silicon nitride film which is stretched onto a quadratic silicon frame. The membrane considered here has dimensions 150 μm ×150 μm ×50nm, a fundamental mode eigenfrequency of $\omega_m = 2\pi \times 0.86$ MHz and an effective mass $m_{eff} = 8 \times 10^{-13}$ kg.

A favourable lattice configuration is a blue detuned lattice where the transverse confinement is provided by a very far detuned 2D lattice. Trapping and Raman sideband cooling of 3×10^8 atoms with a density of 1.1×10^{11} cm^{-3} was demonstrated in [99]. In a 3D lattice with additional cooling the trap loss due to light assisted collisions is reduced as most of the lattice sites are occupied with a single atom [98]. Coupling the mechanical oscillator via the blue detuned lattice to the atoms has the advantage of smaller photon scattering rates in comparison to the red detuned lattice, and thus allows to provide a sufficiently deep potential with a small detuning, small lattice laser beam power, and thus small heating of the membrane. This heating due to laser light absorption can be significant due to the reduced thermal conductivity of the membrane at low temperatures [121]. Reasonable parameters of the blue detuned lattice laser beam are a power of $P = 7$ mW, a waist $w_0 = 230$ μm and a detuning $\delta = 2\pi \times 1$ GHz, which are calculated such that atomic and membrane oscillation frequencies are equal. We assume a spread of detunings of $\Delta\omega_{at} \simeq 2\pi \times 24$ kHz. For the membrane, we assume an amplitude reflectivity of the bare silicon nitride of $R \approx 0.57$ at $\lambda = 780$nm and a mechanical quality factor $Q = 1 \times 10^7$ at $T = 2$ K.

With these parameters, a coherent coupling of $g = 40$ kHz could be achieved, which is large in comparison to the momentum diffusion rate of the membrane $\gamma_m^{diff} = 52$ Hz. The leading decoherence effect of the membrane is due to the coupling to the thermal bath, which contributes a rate $\gamma_m^{th} = \gamma_m \bar{n} = 24$ kHz at 2 K. The diffusion rate of the atomic momentum $\gamma_{at}^{diff} = 16$ kHz is of the order of the cooling rate, which can be tuned up to $\gamma_{at}^{cool} = 20$ kHz [99]. Assuming these optimistic parameters, the hierarchy of the rates is

$$\omega_m = \omega_{at} \gg g \gtrsim \gamma_{at}^{cool} \simeq \gamma_m^{th} \gtrsim \gamma_{at}^{diff} \gg \gamma_m^{diff}. \tag{4.50}$$

4.1 Coupling scheme

In this setting, the action of the atoms onto the membrane can be significant, and one expects sympathetic cooling of the fundamental mode of the membrane oscillator. The steady state phonon number occupation $\bar{n}_{ss}$ is found by solving the Master equation 4.43, and the cooling factor for the assumed parameters is $\bar{n}_{th}/\bar{n}_{ss} \simeq 2\times 10^4$. This would allow for ground state cooling with $\bar{n}_{ss} \simeq 0.8$, starting from a cryogenically precooled membrane at $T = 500$mK.

4.2 Extension of the atom chip setup

The goal of the experiment is to measure the backaction of the atomic ensemble onto the mechanical oscillator, which shows up as an additional damping rate to the oscillators motion in the weak coupling limit. This section describes the setup of an experiment which aims at measuring the additional damping which is given by $\Delta\gamma$, the second term of the equation 4.42.

The setup described in section 3.2 is used as a source of ultracold atoms for loading a 1D optical lattice with a membrane oscillator as an end mirror. The damping of the membrane is measured in a ringdown measurement, where the decay of the initially excited membrane amplitude is observed with a Michelson interferometer. This section focusses onto the setup of a Michelson interferometer, the controlled excitation of the membrane amplitude and the integration of a 1D optical lattice into the atom chip setup described in section 3.2.

4.2.1 Michelson interferometer for membrane readout

The amplitude of the membrane oscillator is read out with a Michelson interferometer as shown in Fig. 4.3. The interferometer consists of a beam splitter (BS) which splits a light beam from a coherent light source (LS) into two beams which are each reflected at a mirror (M1, M2). The reflected beams are overlapped at the beam splitter, and the power is measured with a photodetector (PD).

The power incident on the photodiode depends on the difference of the optical path lengths of the two interferometer arms with optical path lengths L1 and L2. The phase difference is given [122] by

$$\phi = 2\pi(2(L_1 - L_2)/\lambda) \qquad (4.51)$$

where λ is the wavelength of the light. The power incident on the photodetector is given by

$$P = \frac{1}{2}\alpha P_0(1 + C\sin\phi), \qquad (4.52)$$

with the optical power P_0 of the light source. C models the contrast due to unequal powers in L1 and L2. α is an attentuation factor which accounts for optical losses in the interferometer which occur e.g. due to reflections at air-glass interfaces.

The power shows a sine dependence as indicated in Fig. 4.3 (right), if e.g. L_1 is fixed and L_2 continuously varied. The interferometer achieves the highest sensitivity for

4.2 Extension of the atom chip setup

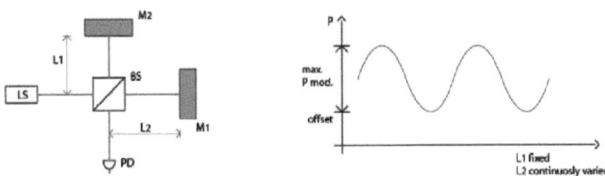

Figure 4.3: (left) Michelson interferometer. (right) Power modulation at the output port for variation of the length of one arm.

an adjustment on the steepest slope. The interferometer measures the amplitude of the membrane. The oscillation amplitude is linearly translated into an oscillating voltage by the photodiode, and can be analyzed with e.g. a Lock-In amplifier.

One fundamental limit to the sensitivity is shotnoise in the photodetector (PD). The poisson distributed shotnoise of a current I has a standard deviation of

$$\sigma_I = \sqrt{\langle i^2 \rangle} = \sqrt{2eI\Delta f}, \qquad (4.53)$$

with the electron charge e and the measurement bandwidth Δf. This has to be compared to the change of the PD current $I + \delta I$ due to a length change of one interferometer arm.

Setup of a Michelson interferometer

The experimental setup of the Michelson interferometer is shown in Fig. 4.4. A home built, free running diode laser at $\lambda = 830$nm is used as a coherent light source. A grating provides feedback into the laserdiode for narrowing the linewidth [123], and is adjusted for single mode operation. The beam is shaped with an anamorphic prism pair (AP) and a telescope, and back reflections from optical components in the beam path into the laser diode are attentuated with an optical Faraday isolator (FI). The laser beam is coupled into a polarization maintaining single mode optical fiber[1] for spatial mode cleaning. Both fiber end facets are angle cleaved to avoid surfaces which are perpendicular to the beam path, and which might lead to unstable power of the transmitted laser beam due to the build up of unrequested Fabry-Perot cavities.

---
[1] Thorlabs PM-780HP

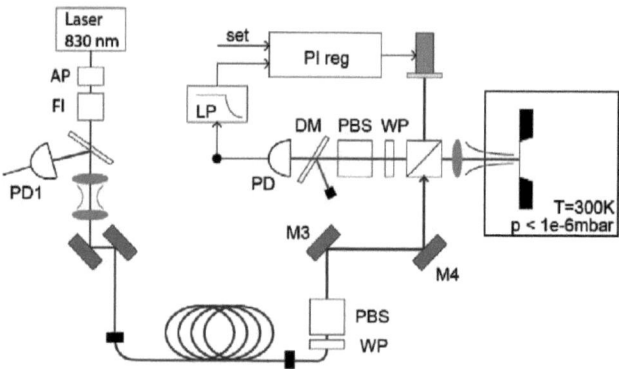

Figure 4.4: Setup of the Michelson interferometer

After outcoupling from the fiber, the beam is collimated and the polarization is cleaned with a $\lambda/2$ plate (WP) and a polarizing beam splitter (PBS) such that the polarization of the transmitted beam is vertically oriented. The power is split at the 50/50 beam splitter in the center of the interferometer, and one light beam is focussed onto the membrane with a lens of $f = 100$ mm. The beam waist on the membrane is approximately $w_0 = 200$ μm. The membrane is mounted on a solid aluminum cuboid in a vacuum chamber which consists of a six way cross (CF40) which is permanently pumped with an ion pump to a pressure $p < 1 \times 10^6$ mbar. A hole at the membrane position in the cuboid and two broadband antireflection coated windows mounted at opposite flanges allow optical access to the membrane from both sides. The membrane frame is UV glued to the cuboid at one corner only in order to avoid bending of the frame due to shrinking of the glue volume during the curing procedure, which might impose unwanted stress onto the membrane. Reflections from the windows into the beam path are avoided by rotating the membrane by $10°$ with respect to the windows. A neutral optical density filter in the reference beam path of the interferometer is introduced to match the powers of the interfering beams. The mirror in the reference arm is a gold coated glass plate which is glued onto a low voltage multi stack piezo, which allows to vary the length of this arm by one wavelength per 10V.

The interfered amplitudes of the overlapped beams are measured with an amplified photo detector (PD) as shown in Fig. 4.5. We use a BPW34 with a reverse bias voltage of 15 V and perform a current to voltage conversion with the operational amplifier OP 37, which combines low noise and a gain-bandwidth product which is sufficient for our purposes. The photodiode circuit provides three different outputs:

4.2 Extension of the atom chip setup

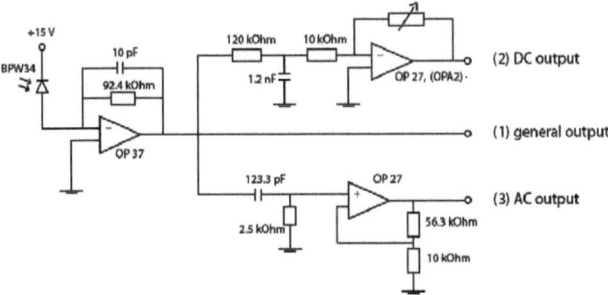

Figure 4.5: Amplification and filtering circuitry of the high sensitivity photo detector.

(1) right after the current to voltage conversion (general out), (2) a low pass filtered signal with subsequent operational amplifier OP 27 (DC out), (3) a high pass filtered signal with subsequent operational amplifier OP 27, providing a gain of 5 to compensate for the drop of the signal level due to filtering (AC out).

To reduce pickup noise it is essential to build the circuit and the photodetector into a massive aluminum box. The wires which guide the current from the photodiode to the current to voltage conversion amplifier should be as short as possible[2]. The adjustment of the resistor in the feedback arm of the operational amplifier follows a trade off between sensitivity and bandwidth. In order to measure the bandwidth of the photodiode, a laser beam which is incident on the photodiode is switched on, and the rise time of the signal at the photodiode is measured at the respective output. The bandwidth is given by the inverse of the time that has elapsed when the signal has reached 63 % of the steady state level. We use the solid state switch in the AOM controller[3] with a switching time of a few tens of ns, which is small in comparison to the timescale on which the signal rises. The bandwidth of the photodiode is measured and adjusted to 1 MHz at the general output. The rms noise at the general output is 3.5×10^{-5} of the full range (12 V).

Operation of the Michelson interferometer

The light beams which are incident on the photodiode are overlapped carefully in order to achieve a maximum modulation of the interfering beams for a variation of

---
[2] A trial to mount the photodiode in a separate box and to guide the current with LEMO connectors and shielded cables to the current to voltage conversion amplifier resulted in significant pick up noise.
[3] AOM_2 100 MHz, Toni Scheich

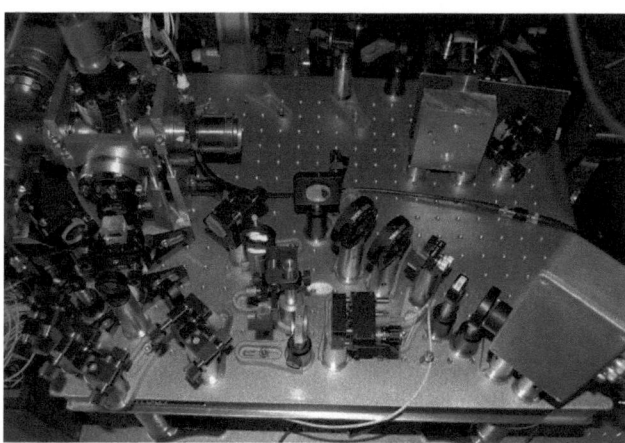

Figure 4.6: Photograph of the Michelson interferometer.

the relative optical path length $(L_2 - L_1)$, and to achieve a minimum power modulation for destructive interference as shown in Fig. 4.3. In the alignment procedure the power of the beam which is reflected from the membrane and coupled back into the optical fiber is monitored with the photodiode (PD1). Maximum power indicates that the beam axis is perpendicular to the membrane, i.e. the incoming and outbound beams are overlapping. A second condition is that the beam hits the center of the membrane. These two requirements are fulfilled simultaneously by aligning the two steering mirrors M3 and M4. In order to overlap the two light beams incident on the photodiode (PD), the power modulation is maximized while tapping the vacuum chamber with the membrane which induces variations of the relative optical path length $(L_2 - L_1)$ of more than $\lambda/2$. The goal of the alignment procedure is that the interference pattern is only one spot, and that the brightness changes common mode over the whole area when $(L_2 - L_1)$ is varied.

Mechanical vibrations which induce arbitrary variations of the relative optical path length $(L_2 - L_1)$ are minimized by choosing stable mechanical components and mounting the interferometer as compact as possible. All relevant optical components are mounted on 1/2 inch posts and in stable holders[4]. The soft spot of the interferometer is the mechanical mounting of the membrane in the vacuum chamber which is susceptible to acoustic vibrations.

---

[4]Maier GmbH

4.2 Extension of the atom chip setup

The relative path length ($L_2 - L_1$) is actively stabilized with a PI regulator[5] which acts on the piezo mirror in the reference arm of the interferometer. To take out acoustic vibrations, it is sufficient to stabilize the interferometer to frequencies in the kHz range. In particular, the stabilization should not work at the frequencies close to the membrane eigenfrequencies. To ensure this, we use the DC out of the photodiode PD and adjust the cutoff of the low pass filter (LP) to 13 kHz. The maximum signal level of the DC output is adjusted with the operational amplifier OPA2 in order to scale to the signal level that the PI regulator requires. The signal from the DC output is galvanically isolated and input as actual value to the PI regulator. The proportional and integral part of the output are summed with an operational amplifier OP 27 in a one to one summing amplifier configuration, and fed to the piezo (PIstab). The set value is adjusted with a potentiometer such that the level of the DC output is stabilized to the steepest slope.

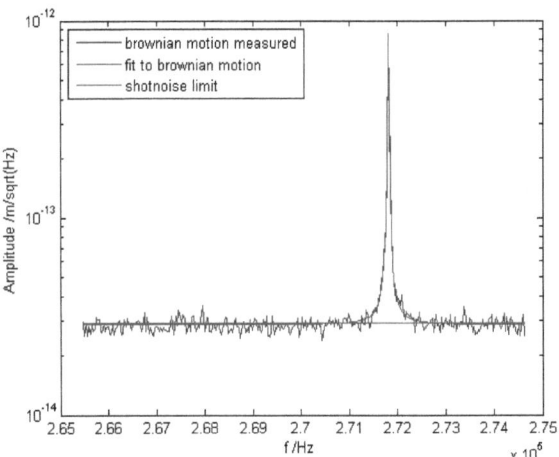

Figure 4.7: The thermal motion of the membrane allows to scale the ordinate.

The interferometer is calibrated with the thermal motion of the fundamental mode of the membrane oscillator which is measured with an oscilloscope[6] at the AC output of the photodiode PD. An FFT is applied to the time trace, and a Lorentzian is fitted to the peak of the power fluctuations of the photodiode current. The integrated area

---
[5]LB$_5$, Toni Scheich
[6]LeCroy waveRunner 44Xi

below the Lorentzian is the squared amplitude of the membrane in units [m^2]. The amplitude is taken as the amplitude of a harmonic oscillator and set equal to $k_B T/2$ in order to calculate a scaling factor for the ordinate. The effective mass of the SiN membrane[7] with dimensions 0.5 mm×0.5 mm×50nm is $m_{eff} = 1.1 \times 10^{-8}$ g. With the calculated rms amplitude of 1.6×10^{-11}m, the ordinate is rescaled as shown in Fig. 4.7 and the noise floor is found to be at 2.9×10^{-14} m/$\sqrt{\text{Hz}}$. A relative length change of $(L_2 - L_1) = 2.9 \times 10^{-14}$m results in a change of the current I of $\delta I = 2.97 \times 10^7$ e/s, if the interferometer is locked to the steepest slope. This coincides with the shotnoise fluctuations of the photodiode current of 70μA measured in a bandwidth of 1 Hz, which is estimated with equation 4.53 to be $\sigma_I = 2.96 \times 10^7$ e/s. This analysis shows, that the performance of the interferometer is limited by shotnoise in the photodetector.

4.2.2 Controlled excitation of the membrane amplitude

Acoustic waves are excited with a low voltage multistack piezo and coupled to the support of the membrane oscillator in order to excite the membrane to a certain amplitude. When the lattice laser is shined onto the membrane the eigenfrequency jitters by several linewidths, as discussed in section 4.3.1 such that stable driving with e.g. a signal generator is not possible. We achieve stable excitation to a set amplitude with a feedback of the interferometer output onto the membrane via a piezo mounted nearby, i.e. through self driving of the membrane in a feedback loop with conrolled gain.

The feedback circuit is shown in Fig. 4.8. The AC output of the photodiode is phase shifted with all-pass filters in order to provide a (frequency) dependent phase shift with unity gain. As one allpass filter shifts the phase by less than 180° in practical implementations, two all-pass filter in series provide all phaseshifts required such that the membrane can be driven resonantly out of the thermal motion. The amplitude of the membrane oscillation is measured with a Lock-In amplifier[8] at the AC output of the photodiode which is also used to lock the local oscillator frequency of the Lock-In. The amplitude of the membrane oscillation is controlled with an integral regulator which acts on a voltage controlled amplifier (VCA). The VCA[9] adjusts the amplitude of the signal that is fed back to the piezo. The integral regulator takes the amplitude output of the Lock-In as actual value and compares it to a set value which is adjusted manually or automatically from the computer control. The integral regulator stabilizes the amplitude of the driving signal via the control

---
[7]Norcada, http://www.norcada.com
[8]SRS SR844
[9]AD602, Analog Devices

4.2 Extension of the atom chip setup

voltage which acts onto the gain of the voltage controlled amplifier. The amplification of the driving signal has a dynamic range from -10 dB to +30 dB.

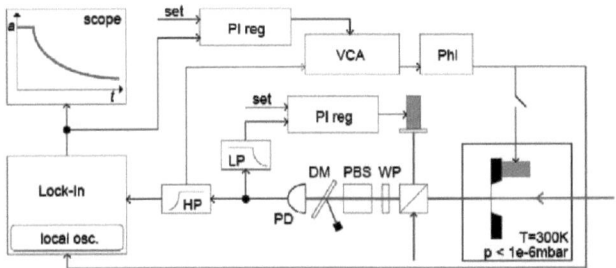

Figure 4.8: Feedback loop for stable membrane excitation.

This home built driving circuitry allows to drive the membrane to any amplitude ranging from the thermal motion up to several nanometers. Different eigenmodes can be selected with the frequency dependent all-pass phase shifters. We use a mechanical relais to close and open the feedback loop.

4.2.3 Optical lattice setup

The atom chip setup described in chapter 3 is used as a source of ultracold atoms which are loaded into a 1D optical lattice potential. The lasers providing the light for the optical lattice are set up on another optical table in the laboratory. A freerunning, grating stabilized master laser[10] seeds a tapered amplifier[11]. The power of the light beam is controlled with an acousto-optic modulator (AOM) which is supplied from an AOM controller[12] and allows to set frequency and power of the radiofrequency signal. A solid state switch allows to switch off the signal within less than 20ns. The light which is diffracted into the first order is coupled to a polarization maintaining single mode optical fiber[13], where both end facets are angle cleaved in order to avoid perpendicular surfaces with respect to the beam path. In addition to spatial mode cleaning, the fiber transfers the beam to the optical table where the experiment is performed. The polarization of the light beam is cleaned after the fiber with a $\lambda/2$ plate and a polarizing beam splitter. The laser beam is focussed

---
[10] DL pro L, Toptica Photonics, with LD-0780-0100-AR-1
[11] BoosTA, Toptica Photonics, with $TA_0 780_0 808$
[12] AOM_2 100 MHz, Toni Scheich
[13] Thorlabs PM-780HP

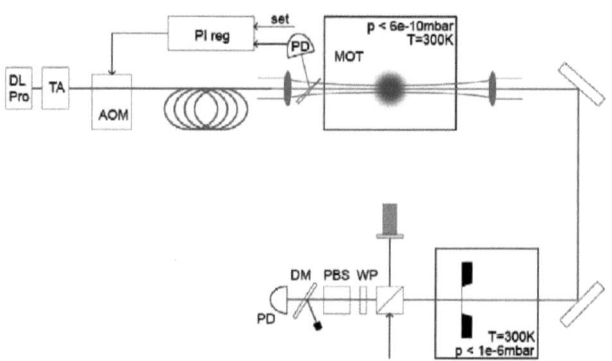

Figure 4.9: Schematic of the optical lattice.

into the atom chip vacuum chamber, and provides a power up to 160 mW.

We use a home built PI regulator for stabilization of the laser beam power. A 4 % reflection is sidelined with a glass wedge which is mounted close to the atom chip vacuum chamber. The power of this beam is measured with a photodiode[14] in order to determine the actual value. The PI regulator compares this with the set value from the computer control and feeds a correction to the modulation input of the AOM controller.

In principle, it would be nice if the bandwidth of the stabilization would exceed the required trap frequencies. However, it is not easy to engineer a PI loop with a bandwidth > 500 kHz with conventional analog components. As the laser is very stable at frequencies > 10 kHz, we restrict the bandwidth of the stabilization to 12 kHz. The output power of the BoosTA varies up to 1 % for frequencies below 10 kHz. In addition, the stability of the laser beam power was improved by filtering the current supply of the BoosTA with capacitors, yielding a total rms noise of 2×10^{-4} of the full level, measured after the optical fiber at a level of $P = 80$ mW.

Every pickup and electronic noise which enters the feedback loop influences the power of the light beam. Similar to the photodiode design it was crucial to build the stabilization circuit and the photodiode into a massive aluminum box to reduce high frequency pickup. As the beam power does not have to be varied quickly in the experiments, the set value voltage from the computer control is filtered with

---
[14]BPW34

4.2 Extension of the atom chip setup

Figure 4.10: Outcoupler of the lattice laser beam.

capacitors. Fast switching for e.g. time of flight measurements is performed with the solid state switch of the AOM controller. Also the power amplifier in the AOM controller is voltage supplied with an external, filtered voltage supply.

The incoming beam from the lattice laser is focussed into the atom vacuum chamber such that the waist is approximately at the position of the MOT. In order to overlap the incoming beam from the lattice laser with the mirror-MOT, we scan the frequency of the laser beam across the atomic resonance. The MOT cloud is imaged with a CCD camera[15]. If the laser beam is on resonance with the atomic transition and hits the cloud, atoms are expelled out of the cloud. The signal is optimized such that maximum overlap of the laser beam with the atom cloud is achieved. The beam is collimated with a lens and focussed onto the membrane with a second lens with $f = 100$ mm to a waist of $w_0 = 250$ μm, and reflected at the membrane. The position of the two lenses is adjusted such that the beam waist of the reflected beam matches the beam waist of the incoming beam in the atom chip vacuum chamber. The length of the optical path between the atom vacuum chamber and the membrane is 0.9m.

In the alignment process, incoming and outbound beam have to be overlapped and should hit the center of the membrane. The first condition is taken into account by coupling the reflected beam to the fiber and monitoring the transmitted power with a photodiode PD, which is a similar procedure as described for the interferometer

---

[15] CCD Mini-Fingerkamera, Conrad

alignment. For the second condition, one makes use of the observation that the frequency of the brownian motion attains a minimum when the beam hits the center and rises monotonously as the beam position is shifted towards the edges, which is due to heating of the membrane by absorption of laser power and the resulting thermal expansion. Both requirements can be fulfilled simultaneously by adjusting the two steering mirrors[16].

If the lattice and the readout laser beam are overlapped, the lattice beam saturates the photodiode (PD) of the Michelson interferometer. The power of the lattice beam is suppressed by making use of the linear polarizations of the readout and the lattice laser beam which are perpendicular to each other. We use a $\lambda/2$ plate, a $\lambda/4$ plate and a polarizing beamsplitter to separate the two beams, and achieve a suppression of 3×10^{-4}. In addition, we use a dichroic beamsplitter[17] which reflects 780nm and transmits 830nm. This beamsplitter attentuates the lattice laser power on the photodiode by another 10^{-2}, such that the total suppression is 3×10^{-6}. This leads to a DC offset of the dark photodiode level by $\simeq 0.7$ V .

---

[16]The aligned lattice laser beam is used to align the interferometer beam to the center of the membrane in a further iteration.
[17]Laserline Dichroic Plate Beamsplitter 64-289, Edmund Optics

4.3 Experimental results

In this chapter, I describe the experimental results achieved with the setup described in the previous section. The main result of this chapter is observation of the interaction of an atomic ensemble with a mechanical membrane oscillator. In section 4.3.3, the membrane is driven to a well-defined amplitude, and excites the COM mode of the atomic ensemble. In section 4.3.4, we observe backaction of the atoms onto the membrane, which shows up as additional damping of the membrane motion. We compare the measurements with the theoretical predictions of the model derived in section 4.1.3, and find agreement between theory and experiment.

4.3.1 Characteristics of the SiN membrane

In our experiment we use quadratically shaped, low-stress SiN membranes[18], with dimensions $(a, a, t) = (0.5 \text{ mm}, 0.5 \text{ mm}, 50 \text{nm})$.

Optical properties

Reflectivity The membrane is a thin dielectric slab with a refractive index of $n_{SiN} = 2.2$, surrounded by vacuum. A light beam impinges perpendicular onto the membrane and is reflected at the first and second surface, as well as between the two parallel surfaces. The intensity reflectivity is given [124] by

$$I_r = I_0 \frac{F \sin^2 \delta/2}{1 + F \sin^2 \delta/2}, \tag{4.54}$$

with $F = \frac{4R_I}{(1-R_I)^2}$, $R_I = \left(\frac{n_{vac}-n_{SiN}}{n_{vac}+n_{SiN}}\right)^2$ and the geometric phase difference $\delta = 2\pi\Delta s/\lambda$, which arises from the optical path difference $\Delta s = 2tn_{SiN}$.

Hence, for SiN membranes with a thickness of $t = 50$nm, on expects an amplitude reflectivity of $R = 0.56$ for $\lambda = 780$nm. The intensity reflectivity of the membrane is measured with a power meter to be $I_r/I_0 = 0.24$, which corresponds to an amplitude reflectivity $|R| = 0.49$. Losses in the optical path reduce the power of the reflected beam further by a transmission $|T| < 1$, and the power of the reflected beam at the position of the atoms is reduced by a factor $|R|^2|T|^4 = 0.176$.

---

[18] purchased from www.norcada.com

Absorption Since the absorption of a SiN membrane is small, it can not be measured precisely with a power meter. One possibility to determine the absorption is to integrate the membrane into a cavity, orient the surface perpendicular to the cavity mode and displace the membrane along the axial direction. Since the membrane thickness t is small compared to λ, this effectively modulates the cavity finesse and allows to infer a value for the absorption. In [58], the absorption of low-stress SiN membranes similar to the ones employed in our experiment is investigated, and the absorption of laser light at $\lambda = 1064$nm in a membrane with $t = 50$nm was found to be $\Im(n_{SiN}) = 1.6 \times 10^{-4}$. In [67], the absorption of high-stress SiN membranes is found to be $\Im(n_{SiN}) \lesssim 10^{-5}$ for laser light at $\lambda = 935$nm.

The absorption of a SiN membrane is in particular important for experiments, where the membrane is kept in a cryogenic environment, as proposed in [38, 39]. The thermal conductivity of low-stress silicon nitride membranes was studied experimentally in [121], and is decreased by a factor of 50 at 1 K in comparison to room temperature. In our room temperature experiment, the membrane temperature is not a critical parameter[19], as our focus is to observe a change in the damping rate.

Mechanical properties

The description of the mechanical eigenmodes of a membrane oscillator was reviewed in section 2.1.1. This section gives a characterization of the mechanical properties of the SiN membrane used in our experiment.

Eigenfrequencies The eigenfrequency of the membrane is measured with the Michelson interferometer as described in section 4.2.1. When we increase the power of the lattice laser beam which impinges on the membrane center, we observe a decrease of the mechanical eigenfrequency.

Fig. 4.11 shows the decrease of the eigenfrequency of the fundamental out of plane mode with increasing lattice laser power P. We attribute the decrease of the eigenfrequency to a relaxation of the tensile stress in the membrane. Most likely, the origin is that the lattice laser beam is partially absorbed in the membrane, and thus leads to an increase of the temperature in the membrane center. This leads to thermal expansion of the membrane, and to a reduction of the tensile stress. We adapt [125] equation 2.15 in order to model the reduction of the membrane eigenfrequency, and assume[20] a heating of the membrane by $\Delta T = 30$K for an impinging power of $P = 100$ mW, based on an absorption of 1.5×10^{-4}. From the equation

---

[19]... as long as the temperature increase does not alter the mechanical properties substantially. From a FEM simulation with typical parameters, one expects an increase of the membrane temperature by several tens of Kelvin.

[20]...from a FEM simulation by Philipp Treutlein.

4.3 Experimental results

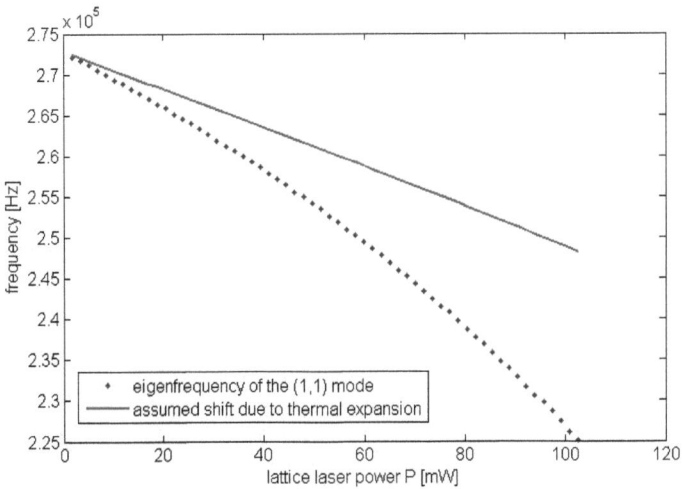

Figure 4.11: The eigenfrequency of the fundamental out of plane mode decreases with increasing lattice laser power. This effect can be caused by heating of the membrane due to absorption of laser light, which yields a relaxation of the tensile stress of the SiN membrane. Comparison with a simple model for the thermal expansion shows, that the eigenfrequency drops more than expected from this model This might be due to higher optical absorption at a laser wavelength $\lambda = 780$nm, in comparison to $\lambda = 1064$nm.

$$\omega_{11} = 2\pi \times \frac{1}{a}\sqrt{\frac{S(1 - \alpha E \Delta T)}{2\rho}}, \qquad (4.55)$$

we calculate the expected shift of the eigenfrequency with the Young's Modulus for SiN $E = 250 \times 10^9$ N/m^2, the density $\rho = 3440$ kg/m^3, and the thermal expansion coefficient $\alpha = 2.6 \times 10^{-6}$ K^{-1}.

The result of this calculation is shown as a line in Fig. 4.11. Comparison with the measurement shows, that the absorption seems to be larger than estimated from the model described above. This indicates, that the absorption at $\lambda = 780$nm is in particular increased with respect to the value reported in [58] for a wavelength $\lambda = 1064$nm.

In order to measure the frequencies of the higher order eigenmodes for various power levels of the lattice laser, we set the power to a constant value. We monitor the time-trace of the AC out of the interferometer (see section 4.2.3 for details), and perform an FFT of the time trace. The thermal motion of each eigenmode shows up as a peak in the spectrum as shown for the fundamental mode in section 4.2.1 in Fig. 4.7. The noise is reduced with averaging, and this allows one to resolve the frequencies of eigenmodes up to 1 MHz.

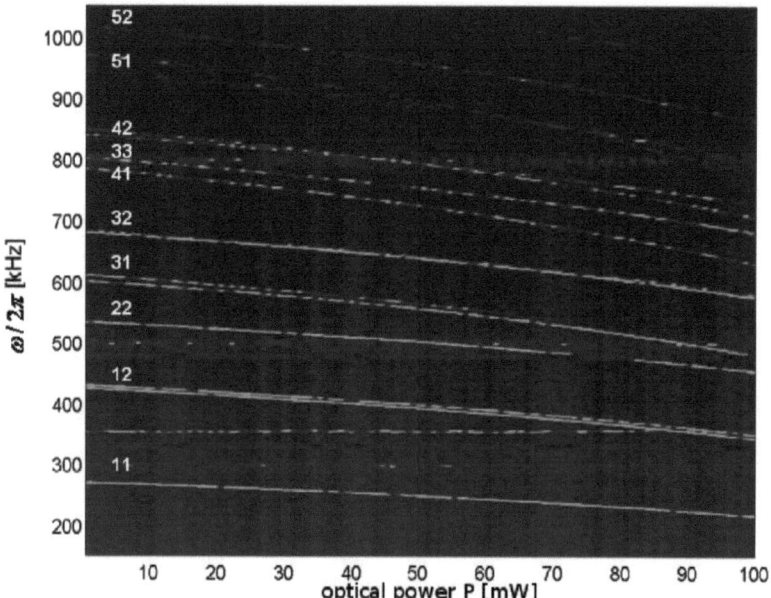

Figure 4.12: Fundamental and higher order mode eigenfrequencies of the SiN membrane. The eigenfrequencies decrease common mode with increasing lattice laser power. The mode indices are indicated in white.

Fig. 4.12 shows the measured eigenfrequencies versus the power of the lattice laser. The lowest line represents the fundamental mode which has a frequency of $\omega_{11}/2\pi = 273$ kHz when the lattice laser is switched off. With equation 2.15, we determine the value of the tensile stress to be $S = 128$ MPa, which is close to the value of $S = 120$ MPa, reported in [58] for a similar membrane with dimensions $(a, a, t) = (1 \text{ mm}, 1 \text{ mm}, 50\text{nm})$. Starting from the value determined for S, one finds

4.3 Experimental results

an agreement between the expected and measured eigenfrequencies of the higher order modes on the percent level. We identify the modes with the following mode indices: (1,1), (1,2), (2,2), (1,3), (3,1), (2,3), (1,4), (3,3), (2,4), (1,5), (5,1), (5,2).

Mechanical quality factor Q The fluctuation of the laser power (for the details of the power stability of the laser, see section 4.2.3) leads to a jitter of the fundamental mode eigenfrequency ω_{11} by several linewidths. This, and the expected [58] small linewidth suggest to measure the mechanical quality factor of the fundamental mode in a ringdown measurement from the $1/e$ decay time τ. In order to achieve a stable initial amplitude, the output of the interferometer is fed back onto the membrane motion via a piezo mounted nearby, as described in section 4.2.2. The mechanical quality factor, measured in the room temperature setup is shown in Fig. 4.13 (top), and found to change with the power of the lattice laser beam.

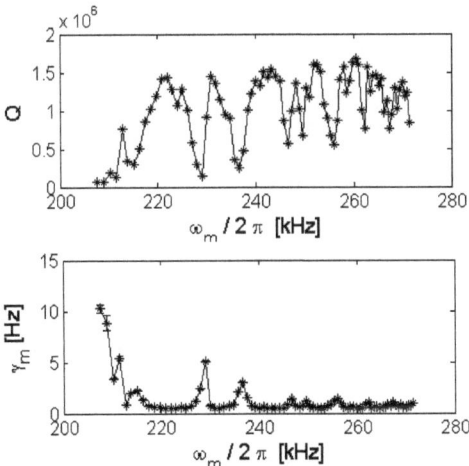

Figure 4.13: (top) Mechanical quality factor of the fundamental mode of the SiN membrane. The pattern is very reproducible, and depends on the actual membrane eigenfrequency ω_m. (bottom) The associated mechanical damping rate γ_m. For the backaction measurements in section 4.3.4, we choose positions where both the mechanical damping and the slope are small in order to achieve comparable conditions.

The behaviour shown in Fig. 4.13 is reproducible, and does not depend on the power, but on the actual frequency of the membrane. The laser power is a handle that allows us to tune the membrane frequency. If we perform the measurement with a different spot size on the membrane, a different dependence of the membrane frequency on the laser power results, but the dependence of the quality factor Q on the membrane frequency is the same. Such a behaviour might arise from coupling of mechanical modes of the membrane to mechanical modes of the support. In [126], the impact of mechanical support modes onto the mechanical quality factor of SiN membranes is studied, and it turns out that the Q-factor for higher harmonics does not decrease monotonically as it would be expected from elastic radiation of energy from the oscillator into the support. The behaviour of the quality factor that we observed might have a similar explanation, and is subject of further research in our group.

4.3.2 Properties of atoms in the optical lattice

The optical lattice is set up as described in section 4.2.3. The linearly polarized 1D optical lattice potential is provided from a grating stabilized diode laser which injects a tapered amplifier. The laser frequency is detuned by $\Delta = -2\pi \times 20.8$ GHz from the D_2 line of ^{87}Rb, and the power P is actively stabilized with a PI regulator in a bandwidth of 12 kHz to compensate for slow drifts; this yields an overall rms noise of 2×10^{-4}. The laser beam with a power ranging from of $P = 0..137$ mW is sent through the MOT chamber and is partially reflected at the surface of the SiN membrane. The incoming and reflected gaussian beam are overlapped and form an optical lattice with a waist of $w_0 = (280 \pm 30)$ μm at the position of the atoms.

The reflected laser beam at the position of the optical lattice is weaker than the incoming beam due to the finite reflectivity of the membrane and losses in the optical path by $|R|^2|T|^4 = 0.18$. This leads to a not fully modulated lattice potential. At a typical power $P = 63.6$ mW, the lattice potential has a total depth[21] of $V_{latt} \approx 250$ μK, if we assume that the beamwaists of the two beams are the same.

We use a mirror-MOT to load the 1D optical lattice potential. The MOT has a magnetic field gradient of 15 Gauss/cm along the optical lattice and all the circular polarized MOT beams are detuned by $\Delta = -2.2\Gamma_{se}$ from the D_2 line. We image the MOT after a short time-of-flight $t_{TOF} = 0.5$ ms with an off-resonantly detuned absorption imaging beam[22], and extract a cloud radius of $\sigma = 190$ μm from a gaussian fit. We assume the density distribution to be symmetric with a density profile given by

---
[21]calculated, $I_{sat} = 16.7$ W/m^2 for π-polarized light on the D_2-line
[22]...in order to avoid saturation of the optical density.

4.3 Experimental results

$$n(r) = \frac{N}{(2\pi)^{3/2}\sigma^3} e^{-\frac{1}{2}\left(\frac{x^2}{\sigma^2}+\frac{y^2}{\sigma^2}+\frac{z^2}{\sigma^2}\right)}, \quad (4.56)$$

and calculate the atomic peak density in the MOT

$$n_p = \frac{N}{(2\pi)^{3/2}\sigma^3}. \quad (4.57)$$

The cooled ensemble, without the lattice being turned on, has a peak density of $n = 2.5 \times 10^{11}$ cm^{-3}, $N = 2.7 \times 10^7$ atoms and a temperature of $T_{MOT} \approx 190$ μK, after a MOT loading time of 6 s.

The optical lattice is loaded by spatial overlapping with the MOT. The lifetimes of the lattice after switching off the MOT are rather short with 24 ms, which might be due to light assisted collisions in our relatively near-resonant lattice, which could be enhanced in a 3D optical lattice, see section 2.2.2. Typical atom numbers in the optical lattice are $N = 0.6..1.5 \times 10^6$ for a lattice with a beamwaist of 280 μm, and scale with w_0^2 for same lattice modulation depths.

In the loading process, we observe a decrease of the temperature by $\Delta T_{MOT} = 30$ μK, and a decrease of the MOT atom number by roughly a factor of 2. The temperature decrease could be due to level shifts introduced by the lattice potential, which might change the effective detuning of the MOT light to larger values. The decrease of the atom number could be due to a competition of the photon scattering rate (equation 2.43) $\Gamma_{sc} \approx 10 \times 10^3$ s^{-1} of the optical lattice[23] with the MOT scattering rate. The total scattering rate of the MOT can be estimated by ignoring possible contributions of the magnetic field, and assuming that the atoms do not remain localized in the standing wave potential created by the MOT beams. In average, the atoms are exposed to light of all polarizations. In this case, the scattering rate[24] is given by [83]

$$\Gamma_{sc,MOT} = \frac{\Gamma_{se}}{2}\left(\frac{I/I_{sat}}{1+4(\Delta/\Gamma_{se})^2+I/I_{sat}}\right). \quad (4.58)$$

With a total power of $P_{MOT} = 14$ mW and an average MOT beam waist $w_0 = 5$ mm, we calculate $\Gamma_{sc,MOT} \approx 6.3 \times 10^6$ s^{-1}.

---
[23]calculated
[24]calculated, $I_{sat} = 35.76$ W/m^2 for isotropic light polarization

4.3.3 Effect of the membrane onto trapped atoms

In a first expriment, we study the effect of the oscillating membrane onto an atomic ensemble trapped in the optical lattice. In order to load atoms to the lattice, the MOT and the lattice are overlapped spatially. In this configuration, absorption imaging of the atomic cloud after a short holding time is not straight forward, since not only atoms in, or released from the optical lattice potential are imaged, but also residual MOT atoms which spoil the absorption images. Due to the short life time of the lattice without the MOT, it is not possible to hold the atoms until the residual MOT have dropped away.

In order to image the atomic ensemble released from the optical lattice after a short time-of-flight (TOF), the repump laser is switched off during the last 1.5 ms of the MOT phase, while the MOT beams remain switched on and transfer MOT and lattice atoms into the state $F = 1$. The lattice beams act as repumper only for the atoms in the lattice volume, which are partially pumped back to the $F = 2$ state. For imaging, we use the cycling transition, such that only atoms which are initially in the $F = 2$ state are imaged. However, in steady state both $F = 1$ and $F = 2$ states are populated at a ratio given by the slow repumping rate of the lattice laser. The ratio is determined by imaging after a long hold time, when the residual MOT atoms have dropped away, with the repumper being switched on. Taking this correction factor into account, the described procedure allows imaging of the atoms trapped in the lattice already after relatively short TOFs $t_{TOF} = 0.5$ ms.

In the experiment, the membrane is continuously driven at a fixed amplitude of $d = 325$ pm during the whole experimental cycle. The atomic ensemble is hold for $t_h = 5$ ms, after the MOT is switched off, before it is released into the TOF by switching off the lattice within 20ns. The temperature of the atomic ensemble is determined from TOF measurements of the width of the atomic ensemble for the case when the membrane is driven to $d = 325$ pm, and compared to the undriven case with a thermal amplitude of $d = 12$ pm.

In order to resolve the effect spectrally, the trap frequency is varied by scanning the lattice laser power P. From these measurements we observe an increase of the temperature of the atomic ensemble along the optical lattice as shown in Fig. 4.14. The axial and radial temperatures are respectively normalized to reference measurements without membrane driving, and the ordinate shows the increase of the temperature due to the membrane oscillation.

The frequency axis is calculated with equations 2.41 and 2.45 from the lattice parameters. We use the measured detuning $\Delta = -2\pi \times 20.8$ GHz, the waist

4.3 Experimental results

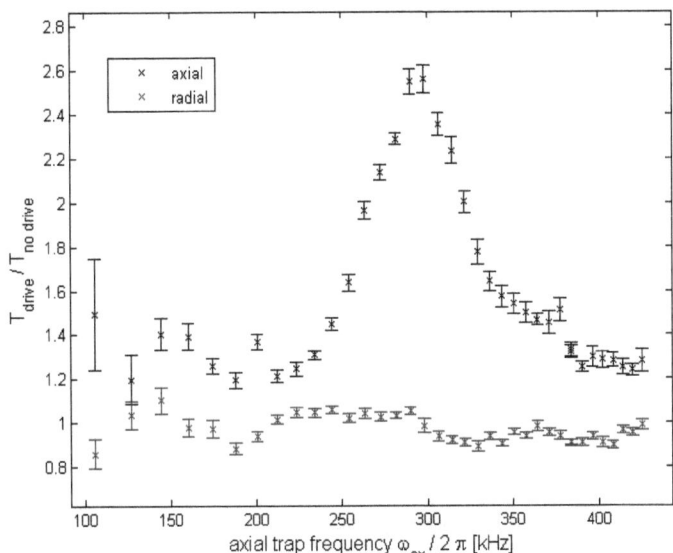

Figure 4.14: Spectroscopy of atoms trapped in the lattice. The temperature of the ensemble is normalized to the temperatures without driving of the membrane in order to take out systematic heating effects due to e.g. photon scattering. Resonant driving affects the temperature along the lattice direction, whereas the temperature along the radial direction remains nearly unchanged. At the center of the resonance, the membrane oscillates at a frequency of $\omega_m = 2\pi \times 240$ kHz, and the peak appears at higher trap frequencies than expected. The FWHM of the resonance is $2\pi \times 50$ kHz

$w_0 = 280$ µm, and the total amplitude reflectivity[25] of $|R||T|^2 = 0.42$. For the power $P = 63.6$ mW at the position of the peak, we calculate the axial trap frequency $\omega_{ax} = 2\pi \times 283$ kHz. This is, however in contrast to the expectations from the experimental situation, since the membrane oscillates with a frequency of $\omega_m = 2\pi \times 240$ kHz when the power of $P = 63.6$ mW is impinging. Therefore, one would expect the peak of the resonance to be around $\omega_{ax} = 2\pi \times 240$ kHz.

This discrepancy is within the errorbars of the values which enter into the calculation. Already the uncertainty of the beamwaist $w_0 = (280 \pm 30)$ µm gives rise to an error of 10 percent. Also the assumption, that the beamwaists of the incoming and reflected beam are the same is not true. An analysis of the reflected beam with a beam profiler reveals, that the transverse beam profile is strongly distorted and shows a double peak structure[26]. The uncertainty of the beam waist of the reflected beam is more like $w_0 = (280 \pm 100)$ µm. Already this would explain the discrepancy of measured and calculated trap frequency.

In the following, we discuss two other sources which give rise to a systematic shift. First, the atoms do not oscillate in a harmonic, but in a $\cos^2$ shaped potential. The anharmonicity of the potential shifts the effective oscillation frequency of an atom towards a smaller frequency. In addition, the resonance is distorted by the shift of the membrane frequency with laser power. Second, the trap frequency varies spatially in the lattice volume with the intensity, which changes due to the gaussian beam profile in the radial direction, and due to the divergence of the gaussian beam in the axial direction.

Shift of the trap frequency due to anharmonicities The trapping potential is harmonic only to first order, and for large excursions of an atom in the trap, the anharmonicity of the $\cos^2$ potential increases the oscillation period. This yields an effective oscillation frequency of an atom, which is smaller than the trap frequency calculated for a harmonic potential with equation 2.45. We perform a simulation of the classical trajectory of an atom in the $\cos^2$ shaped potential. For the amplitude of the oscillation, we calculate the axial radius of the thermal ensemble, which has a temperature of $T = 80$ µK at the peak position, if the membrane is not driven[27]. We assume that the axial oscillation amplitude is equal to the thermal radius, multiplied by the square root of the temperature increase that we observe in Fig. 4.14 in the situation, where the membrane is driven. This analysis shows that the effective oscillation frequency of an atom can be shifted by a factor of 0.92 with respect to

---

[25]...also taking losses in the optical path between atoms and membrane into account.
[26]Private message from Maria Korppi
[27]The temperature of the trapped ensemble is smaller than the temperature of the MOT due to thermal evaporation

4.3 Experimental results

the trap frequency calculated with equation 2.45. For example, the trap frequency at the peak position is reduced from $\omega_{at} = 2\pi \times 283$ kHz to $\omega_{at} = 2\pi \times 260$ kHz due to the anharmonicity. This shows, that anharmonicity can have a significant impact onto the position of the resonance.

Shift of the membrane oscillation frequency with laser power Another effect which has to be taken into account is the dependence of the membrane eigenfrequency on the lattice laser power which is discussed in section 4.3.1. The membrane eigenfrequency is shifted, when the trap frequency is scanned by variation of the lattice laser power. This shift is directed in the opposite direction compared to the change of the trap frequency. For example, if the lattice laser power is increased, the membrane eigenfrequency drops. Hence, the resonance appears to be too narrow.

We calculate the anharmonic frequency shift for each data point of the data shown in Fig. 4.14. Fig. 4.15 shows the same data, plotted over the effective oscillation frequency of an atom which is extracted from the model discussed above. The peak of the resonance is shifted towards lower frequencies.

Classical model for the spread of detunings In this paragraph, we develop a classical model based on the spread of oscillation frequencies in the gaussian lattice geometry.

- The transverse spread of detunings arises from the transversally shaped gaussian beam profile. We simulate the response of an atom in a $\cos^2$ shaped trapping potential to a modulation of the trap minimum position. Such response functions for different ω_{ax} are weighted according to the number of atoms which actually oscillate at the respective trap frequency, which is numerically determined from the thermal distribution in the transverse lattice potential. The model reproduces the width of the resonance as shown in Fig. 4.15.

- The axial spread of detunings arises from the axially decreasing maximum intensity of individual lattice sites from the waist position along the optical lattice. The waist of $w_0 = 280$ μm implies a Rayleigh range of $z_{ray} = \pi w_0^2/\lambda = 0.32$ m, which renders the axial spread of detunings negligible: the highest trap frequency of an individual site drops by $\Delta\omega = 2\pi \times 0.35$ kHz at the radius $\sigma = 190$ μm of the MOT used for loading.

In conclusion, the discrepancy between measured and calculated trap frequencies can be explained with uncertainties of the beamwaists of incoming and reflected

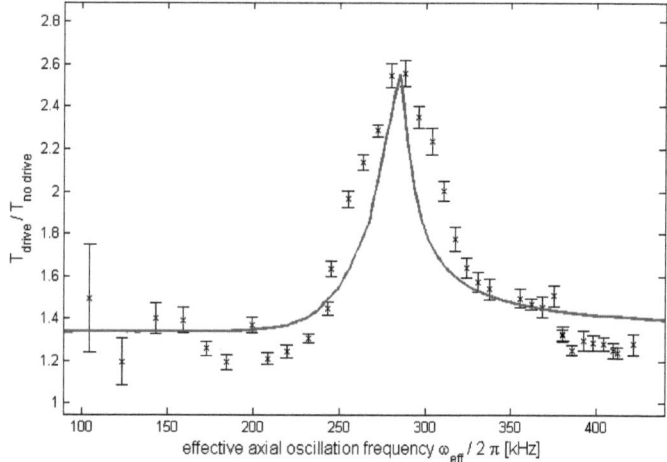

Figure 4.15: In comparison to Fig. 4.14, the abscissa shows the effective oscillation frequency of a trapped atom in a $\cos^2$ shaped potential at an amplitude corresponding to the thermal radius of the ensemble. The anharmonicity of the potential shifts the effective oscillation frequency of an atom towards a smaller frequency. The resonance (continuous line) is expected to coincide with the membrane eigenfrequency $\omega_m = 2\pi \times 240$ kHz. The prediction (line) from of a classical model based on the transverse spread of trap frequencies shows that this effect could explain the shape of the resonance. The amplitude and the frequency offset of the predicted line are fitted to the data.

4.3 Experimental results

beams. Anharmonicities and spread of detunings yield a shift into the right direction; however, these shifts are too small to be considered as an exclusive explanation.

4.3.4 Backaction of the atoms onto the membrane

In this section, we investigate the backaction of the atoms onto the membrane which shows up as an additional damping rate in equation 4.42, which is derived in the theory section 4.1.3. Due to the sub-hertz linewidth, we determine the quality factor Q from amplitude ringdown measurements, where the $1/e$ decay time of the amplitude is measured.

In principle, the interaction of the membrane with the atomic ensemble would be limited by the lifetime of the atomic ensemble in the optical lattice to several tens of milliseconds[28] To realize an interaction on a longer time scale, the MOT is permanently replenishing the optical lattice during the ringdown measurement such that a constant number of atoms is trapped in the optical lattice.

A typical measurement sequence is as follows. The optical lattice is permanently loaded during the experimental cycle with atoms from the overlapping MOT which runs in steady state. The membrane is driven to a fixed amplitude of $d = 540$ pm with a feedback of the interferometer output onto a piezo as described in section 4.2.2. When the feedback loop is opened, the membrane amplitude decays exponentially, and is monitored on the scope.

As discussed in section 4.3.1, the quality factor, and therefore the ringdown time constants depend on the actual frequency of the membrane. In particular, decay times of ringdowns which are taken at a membrane frequency where the derivative $d\gamma_m/d\omega_m$ is large (see Fig. 4.13 (bottom)), show a significant scatter in subsequent experiments. Also ringdowns taken at membrane frequencies ω_m where γ_m is large show a significant scatter of the decay times. In order to ensure comparable measurement conditions in subsequent measurements, measurements are only taken at points, where γ_m is small.

We perform alternating experiments with and without atoms in the lattice[29], and determine the respective amplitude decay rates Γ and $\gamma_m/2$ from an exponential fit to averaged measurements. This allows us to determine the change of the decay rate due to the presence of the atoms $\Delta\gamma$. The statistical errors are determined from the analysis of subsets of the collected data.

---

[28]Possible reasons for this short lifetime are light assisted collisions, which could be suppressed in 3D optical lattices, see section 2.2.2.
[29]The presence of atoms is controlled by detuning the frequency of the MOT lasers.

We perform two different experiments. First, we scan the trap frequency as in the previous section in order to resolve the coupling spectrally. In a second experiment, we prepare the atoms at a fixed trap frequency and vary the atom number. This measurement allows us to make a comparison with theoretical predications.

Resonance of the backaction

Similar as in the experiment described in section 4.3.3, we vary the power P of the lattice laser beam in order to resolve the strength of the backaction spectrally. For each power level we take 455 ringdown measurements, and alternate measurements with and without atoms in the optical lattice. The raw data is respectively averaged, and we extract the ringdown time constants from an exponential fit which allows us to determine the additional membrane damping $\Delta\gamma$. This procedure is only valid as long as there are no significant drifts, what is checked by evaluating the Allan variance of the data sets. Since the signal is not too far above the noise level, it is important that the drift of the power of the lattice laser and the readout sensitivity is minimized. The details of the stabilization of the lattice laser power and the interferometer are discussed in section 4.2.

The result is shown in Fig. 4.16. The frequency axis is calculated from the analytical formula, i.e. without taking anharmonicities into account. We observe a broad resonance, which seems to consist of two peaks. The double peak structure could possibly arise from the distorted beam profile of the reflected beam. The resonance (light grey crosses) seems to be centered around $f_0 \approx 280$ kHz in Fig. 4.16, and is shifted by 40 kHz in comparison with the result from the previous section. If one considers only the peak at $f_0 \approx 280$ kHz, the width seems to be similar to that resonance.

Backaction vs. atom number

In order to compare the coupling mechanism more quantitatively to theoretical predictions, the atomic ensemble is prepared at the maximum of the backaction. The atom number is controlled by varying the repump laser power or detuning, and measured as described in section 4.3.3. Fig. 4.17 shows the measured decay of the energy $\Delta\gamma$. We observe a linear scaling of $\Delta\gamma$ with the atom number N, as expected from the theory.

The theory predicts, that the additional (energy) damping $\Delta\gamma$ due to the atoms scales with the atom number N as (see equation 4.41),

$$\Delta\gamma = \frac{\omega_{at}^2 |R|^2 |T|^2 N m}{\gamma_{at} M}. \tag{4.59}$$

4.3 Experimental results

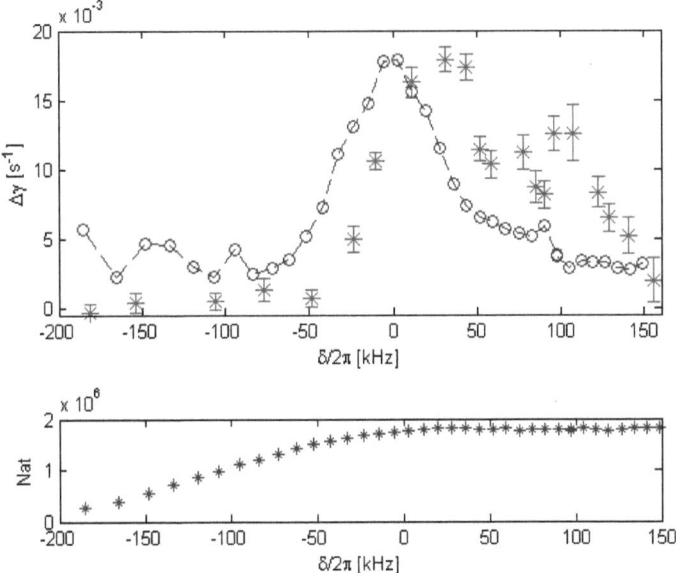

Figure 4.16: (top) Spectrum of the backaction of the atoms onto the mechanical oscillator. The abscissa shows the detuning from the peak position of the data analyzed in Fig. 4.14 (dark grey circles). The resonance of the backaction is shifted to higher frequencies. Further, the resonance is significantly broader than expected from Fig. 4.15, and shows a reproducible double peak structure. The double peak structure might arise from a transversally distorted trapping potential, which might arise from a distorted beam profile of the reflected beam, which was observed under a certain alignment of the lattice configuration. (bottom) Atom number during the coupling experiment.

We extract the dissipation rate of the atoms γ_{at} from the width of the backaction resonance shown in Fig. 4.16. For an estimate, we take a value of $2\pi \times 50$ kHz for the FWHM, read off for the peak centered around 280 kHz. We calculate $\Delta\gamma$ according to equation 4.59 for the mass m of an ^{87}Rb atom, the effective membrane mass $M = 1.1 \times 10^{-8}$ g, and the reduction of the coupling constant g due to optical losses $|R|^2|T|^2 = 0.21$. For $\gamma_{at} = 2\pi \times 50$ kHz and $N = 1.5 \times 10^6$ atoms, the theory predicts $\Delta\gamma = 0.0142$ s^{-1}. The interpolation of the measured value $\Delta\gamma = 0.0115$ s^{-1} at this atom number agrees very well with this theoretical prediction.

Optomechanical coupling via an optical lattice

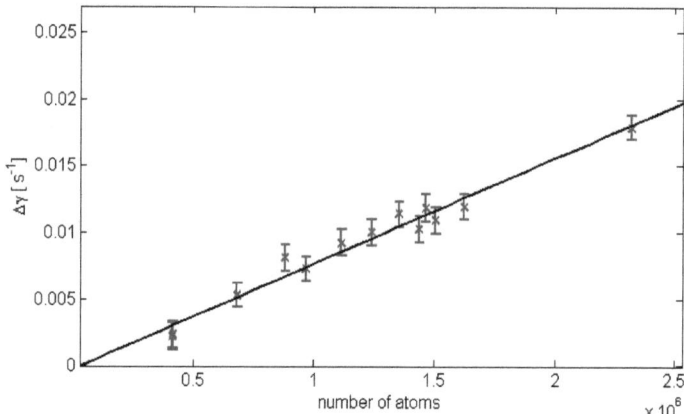

Figure 4.17: Additional membrane damping $\Delta\gamma$ due to the coupling to the atomic ensemble. The straight line is a linear fit.

The uncertainties in this equation arise mainly from the atom number and γ_{at}. Moreover, due to the width of the resonance in Fig. 4.16, and its double peak structure, it is uncertain how many atoms actually contribute to the coupling. In principle, the atom number can be determined precisely by absorption imaging. In our case, we are limited by the short trap lifetimes of $\tau_{trap} \approx 24$ ms, which makes it impossible to take the absorption image after the residual MOT atoms have dropped due to gravity. For short TOFs, the results for the atom number are influenced by the background, which is generated by MOT atoms which are not residing in the dark state $F = 1$. However, our imaging method as described in section 4.3.3 allows to make a relative comparison of the trapped atom number in subsequent shots of the experiment, such that we can observe the linear scaling in Fig. 4.17.

In conclusion, we demonstrated the backaction of an atomic ensemble onto a membrane oscillator for the first time. The results agree nicely with the expected linear scaling of the backaction with the atom number, and the magnitude of the measured backaction coincides with the theoretical expectations. The absolute strength differs by a factor $1/4$ from the theoretically expected value, which may be explained from large uncertainties in the determination of the absolute atom number and the width of the backaction resonance, which shows a reproducible double peak structure.

5 Coupling to the collective spin of a BEC

In the experiments described in chapters 3 and 4, the motion of the mechanical oscillator is coupled to the center of mass motion of ultracold atoms. Coupling to the collective spin of a Bose-Einstein condensate (BEC) bears the advantage that the coupling strength is no longer limited by the square root of the mass ratio. Moreover, the atomic spin typically has a significantly higher coherence time [26] in comparison to the center of mass (COM) motion which couples to other modes in the anharmonic trapping potential.

In the following, I review the coupling mechanism and give details of a possible experimental implementation. With optimistic parameters the strong coupling regime seems to be within reach. These results are published in [31]. Moreover, I will report on the status of a fabrication process which was developed in order to realize the envisaged structure. Finally, I propose a simplified method to fabricate nano-sized mechanical oscillators which are functionalized at their tip with a single-domain magnet.

The fabrication was carried out in the cleanroom facilities of the nanophysics group of Prof. Kotthaus at the LMU München. Without his generous allowance, to access the very well equipped cleanroom, the experimental results presented in this chapter would not have been possible.

5.1 Coupling scheme

A mechanical oscillator with an out-of-plane fundamental mode is functionalized with a single domain cuboid shaped magnet which creates a magnetic field, and an associated field gradient G_m. The physical situation is illustrated in Fig. 5.1. The fundamental mode oscillations $a(t) = a\cos(\omega_r t + \phi)$ are transduced into an oscillating magnetic field $\mathbf{B}_r(t) = G_m a(t)\mathbf{e}_x$ at a distance y_0, $y_0 \gg a(t)$. ^{87}Rb atoms are trapped in a Ioffe-Pritchard type magnetic trapping potential at a distance $y_0 \approx 1\mu m$ above the cantilever in spin states $|F, m_F\rangle = |2, 2\rangle$ or $|F, m_F\rangle = |1, -1\rangle$. The energy splitting between neighbouring m_F levels is given by the Larmor frequency $\omega_L = \mu_B |g_F| B_0/\hbar$. $\mathbf{B}_r$ is perpendicular to the static field $\mathbf{B}_0 = B_0 \mathbf{e}_z$ in the magnetic

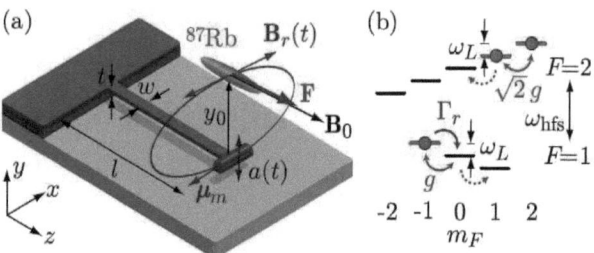

Figure 5.1: BEC-mechanical oscillator coupling mechanism. (a) Atom chip with a BEC of ^{87}Rb atoms (red: BEC wave function) at a distance y_0 from the mechanical oscillator. The freestanding structure (dark grey) is supported at one end and performs out-of-plane mechanical oscillations $a(t)$. A single-domain ferromagnet (purple) on the oscillators' tip creates a magnetic field with oscillatory component $\mathbf{B}_r(t)$ which couples to the atomic spin $\mathbf{F}$. (b) Hyperfine structure of ^{87}Rb in the magnetic field $\mathbf{B}_0$. Transitions between the hyperfine levels $|F, m_F\rangle$ are driven (arrows), if the Larmor frequency ω_L is tuned to the oscillation frequency of the mechanical oscillator. Magnetically trappable states are indicated.

trap. The interaction of the atomic spin $\mathbf{F}$ and the mechanical oscillator is given by the Zeeman Hamiltonian

$$H_Z = -\boldsymbol{\mu} \cdot \mathbf{B}_r(t) = \mu_B g_F F_x G_m a(t) \tag{5.1}$$

where $\boldsymbol{\mu} = -\mu_B g_F \mathbf{F}$ is the operator of the atomic magnetic moment.

The Larmor frequency can be tuned in the MHz range by adjusting the static magnetic field in the trap center. Resonant coupling can be switched on and off by controlling the detuning $\delta = \omega_r - \omega_L$. In the case of resonant coupling ($\delta = 0$), the mechanical oscillators' motion drives spin flips to untrapped internal states $|F, m_F\rangle = |2, 0\rangle$ or $|F, m_F\rangle = |1, 0\rangle$, with an associated continuum of motional states. The coupling shows up as an additional loss rate Γ_r of trapped atoms.

The situation is analogous to a continuous wave atom laser, where a radio frequency field couples trapped atoms via spin flip transitions to a continuum of untrapped motional states. Fig. 5.2 shows the harmonic trapping potential for a BEC in the Thomas-Fermi (TF) limit. The parabola shaped deformation of the energy of untrapped motional states is due to the repulsive potential generated by the mean field of trapped atoms and reflects their density distribution. Hence, the transition

5.2 Design for achieving a strong coupling

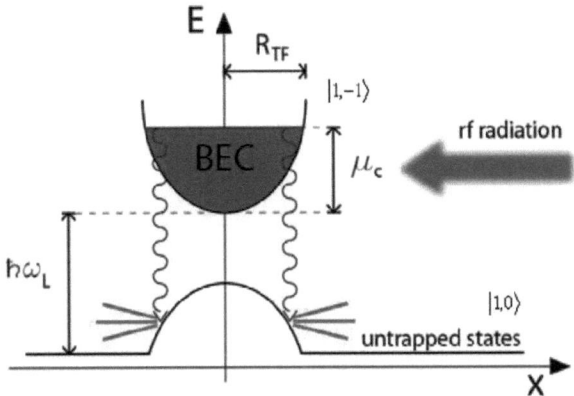

Figure 5.2: The action of the mechanical oscillator onto the atoms is similar to an atom laser. In our case, spin flips of trapped atoms to untrapped motional states are driven from an oscillating magnetic field which is due to the oscillator motion. The transition is broadened by the chemical potential μ_c.

is broadenend by the chemical potential μ_c.

To derive Γ_r, one follows the description of an atom laser given in [127]. The approach given there includes atomic interactions and neglects gravity. This is justified due to the high trap frequencies of approximately 10 kHz. A trapped BEC in state $|F, m_F\rangle = |1, -1\rangle$ in the TF regime is coupled to untrapped states $|F, m_F\rangle = |1, 0\rangle$ with Rabi frequency $\Omega_R = \mu_B G_m a/\sqrt{8}\hbar$. For typical parameters and $\hbar\Omega_R \ll \mu_c$, only a fraction $\simeq \hbar\Omega_R/\mu_c$ is coupled out of the BEC with a rate

$$\Gamma_r = \frac{15\pi}{8} \frac{\hbar\Omega_R^2}{\mu_c}(r_c - r_c^3), \tag{5.2}$$

with $r_c = \sqrt{\hbar\delta/\mu_c}$. Only a fraction $\hbar\Omega_R/\mu_c$ of the trapped atoms on the ellipsoidal shell with main axes $r_i = r_c R_{TF,i}$ fulfills $\delta = 0$, and is coupled to the continuum of untrapped motional states.

5.2 Design for achieving a strong coupling

In this chapter we discuss how to implement the coupling scheme, and give design guidelines for the experimental realization. The goal is to read out the thermal motion of a mechanical oscillator in a room temperature environment.

The Rabi frequency $\Omega_R = \mu_B G_m a/\sqrt{8}\hbar$ scales linearly with the amplitude a of the mechanical oscillation and with the gradient G_m of the magnetic field at the position of the trapped atoms; to achieve a strong coupling, the thermal amplitude x_{rms} and G_m have to be maximized.

5.2.1 Large thermal amplitude x_{rms}

The amplitude x_{rms} of the thermal motion of a mechanical oscillator is

$$x_{rms} = \sqrt{\frac{k_B T}{m_{eff}\omega_r^2}} \tag{5.3}$$

To obtain a large amplitude it is favourable to minimize the product $m_{eff}\omega_r^2$. The mechanical frequency is also subject to another condition: it has to be matched to the Larmor frequency of the trapped atoms in order to enable resonant coupling, and should therefore be in the MHz range. The fundamental mode eigenfrequency of a mechanical oscillator clamped at one side to the support is given by

$$\omega_r \approx 2\pi \times 0.16 \sqrt{\frac{E}{\rho(1+c)}} \frac{t}{l^2} \tag{5.4}$$

with Young's Modulus E, density ρ, length l, width w and thickness t. $c = m/0.24\rho lwt$ accounts for an additional mass at the tip of the mechanical oscillator. To minimize the product $m_{eff}\omega_r^2$, the mass and therefore the dimensions of the mechanical oscillator should be as small as possible. As the frequency scales with t/l^2, it is favourable to increase the length l and reduce width w and thickness t to achieve a small effective mass m_{eff} at fixed ω_r.

Mechanical oscillators out of silicon can be fabricated with typical dimensions of a few hundred nm to several micrometers. Experimentally, we found that such mechanical oscillators with dimensions $(l, w, t) = 7.5$ μm$\times$250nm$\times$100nm can be reproducibly fabricated. Fig. 5.3 shows an array of single side clamped mechanical oscillators.

5.2.2 Large magnetic field gradient G_m

In order to generate the magnetic field (gradient), the tip of the mechanical oscillator is functionalized with a permanent magnet. To optimize Γ_r, the magnet should

5.2 Design for achieving a strong coupling

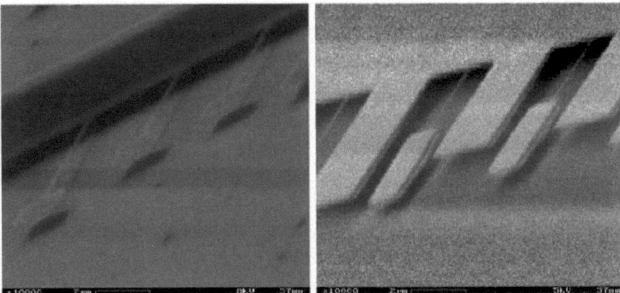

Figure 5.3: Single side clamped mechanical oscillators. (left) Single side clamped mechanical oscillators can be reproducibly fabricated up to a length of 7.5 μm. The eigenfrequency of such oscillators can be matched with the Larmor frequency of trapped atoms. (right) A gold mirror at the tip of the mechanical oscillator allows to readout the mechanical motion interferometrically.

have a small mass and exhibit a strong magnetic field gradient. The magnetic properties of the magnet should not be affected during the coupling experiment, when the magnet is exposed to the magnetic fields which provide the magnetic trapping potential for the atoms. Furthermore, it should be possible to define the shape of the magnet lithographically and to deposit the material with a standard deposition process.

If the size of a magnetic structure is of the order of several magnetic domains, the shape determines the pattern of the domain structure, and a configuration of minimal potential energy is favourable. Reducing the size of the magnetic structure can further lead to the formation of single domain magnets which consist of a single magnetic domain. Choosing an appropriate shape allows to determine the direction of the magnetization. A favourable geometry is a cuboid where the long axis exceeds the dimension of the cross-sections. The magnetization is then aligned along the long axis. The field that needs to be applied to reverse the magnetization is called switching field. Fig. 5.4 shows a comparison of the hysteresis loops of a single domain magnet [128] and a macroscopic ferromagnet [129] which consists of a large number of domains.

Single domain magnets have a high remanent magnetization because all elementary magnets are part of the same magnetic domain, and this is the optimal configuration to realize both a small mass of the magnet and a strong magnetic field gradient. In comparison to other magnetic materials like e.g. nickel, the switching field of a cobalt single domain magnet exceeds 500 Gauss [130] for magnet lengths $l_m > 1$ μm. This

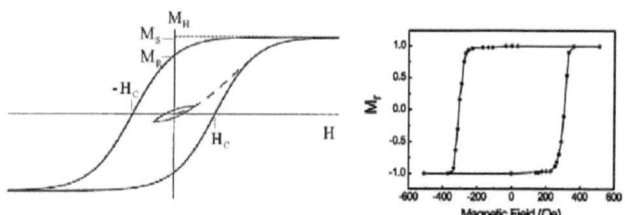

Figure 5.4: (left) Hysteresis loop of a macroscopic ferromagnet [129] and reconstruction of the hysteresis loop of a single domain magnet [128] (right).

is larger than the typical magnetic fields of 100 Gauss which are used for magnetic trapping. Simulation of the magnetic gradient field provided by a bar shaped cuboid [131] shows that the maximum of the magnetic field gradient is at the distance y_0 if $l_m = y_0$. As the atoms can be trapped at a distance $y_0 \approx 1$ µm from the magnet, we choose $l_m = 1.2$ µm. We measured a switching field of 200 Gauss [132] for a single domain magnet with dimensions $(l_m, w_m, h_m) = (1.1$ µm, 225nm, 70nm$)$.

5.2.3 Distortion of the magnetic trap

The field gradient of a magnetic dipole is given by $G_m = 3\mu_0|\boldsymbol{\mu}_m|/4\pi y_0^4$ in the geometry of Fig. 5.1. The gradient depends linearly on the magnetic moment $\boldsymbol{\mu}_m$, and for a given distance y_0, the equation suggests that Γ_r can be increased by simply increasing the magnetic moment $\boldsymbol{\mu}_m$. However, this is not possible, as the trapping potential for the atoms is strongly distorted by this additional magnetic field gradient.

To reduce the impact on the magnetic trapping potential, the magnet on the mechanical oscillator can be enclosed with two bar magnets of same cross section at both ends. The magnetization of the three magnets is aligned parallel. The two long magnets guide the magnetic field of the center magnet which is sitting on the mechanical oscillator away such that the static magnetic field gradient at the position of the trapping potential is effectively reduced. However, the dynamic field gradient which mediates the coupling is not affected.

Fig. 5.5 (a) shows the magnetic stray field of a single domain magnet and how this is modified, when two enclosing magnets are added. The BEC is trapped at a distance y_0. Along a line (red dashed line in Fig. 5.5 (c)) at distance y_0 parallel to the long axis of the magnets, the gradient of the magnetic field component $\partial B_x/\partial y$ is evaluated. This is the relevant component for coupling to the spin of trapped atoms. The enclosing magnets change the magnetic field distribution, and hence

5.3 Resolving thermal motion

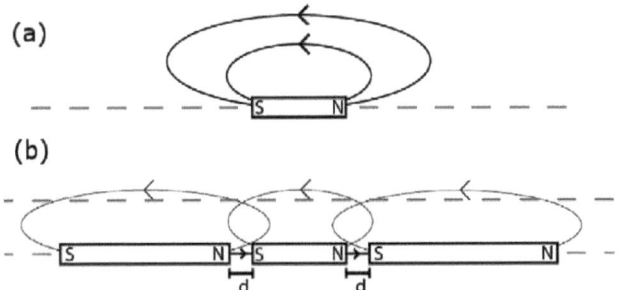

Figure 5.5: (a) Magnetic stray field of a single domain magnet. (b) Two magnets enclosing a single domain magnet reduce the magnetic field gradient along the (displaced) dashed line.

the gradient. Fig. 5.6 shows the magnetic field gradient calculated [131] along the displaced dashed line in Fig. 5.5 (b). The dark grey line shows the magnetic field gradient without enclosing magnets. The other lines show the effect, when enclosing magnets of 9 μm length and same cross section as the center magnet are added, with a gap between the center magnet and enclosing magnets of $d = $ 200nm and $d = $ 50nm. This shows, that the gradient can be reduced considerably by varying the gap d between the center magnet and the two enclosing magnets.

5.2.4 Trapping of atoms

In chapter 3, the trapping of ultracold atoms and Bose-Einstein condensation on an atom chip is described. Instead of the AFM cantilever chip used in that experiment, an additional chip which contains the mechanical and magnetic structures as discussed in the previous sections (Mechanical oscillator chip, Mo chip) is fabricated in order to be glued on top of the atom chip package. MOT and molasses cooling, as well as loading to a first magnetic trap on the chip and transport in the waveguide can be realized in the same way. After the transport to the Mo chip the atom cloud can be loaded to a Ioffe-Pritchard trap which is created with wires close to the mechanical oscillator and condensed with radio frequency evaporation.

5.3 Resolving thermal motion

The considerations in section 5.2 lead to the layout shown in Fig. 5.7 (a). The single side clamped mechanical oscillator out of silicon has dimensions $(l, w, t) =$

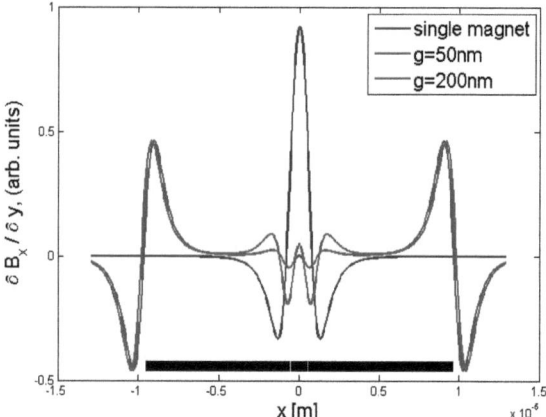

Figure 5.6: The magnetic field gradient is reduced with additional compensation magnets. The simulation shows the magnetic field gradient of the center magnet evaluated along the dashed line in Fig.5.5 (b) without enclosing magnets (dark grey line), and enclosing magnets with at a gap distance $d = 50$nm and $d = 200$nm (light grey line). The black bars indicate the position of the magnets.

5.3 Resolving thermal motion

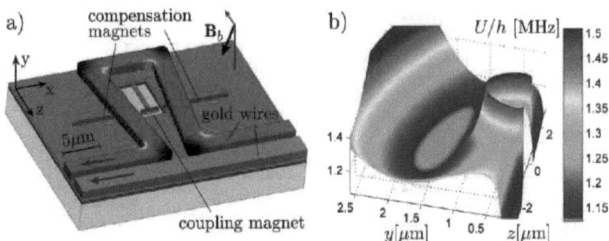

Figure 5.7: (a) Mo chip layout. The trapping potential is created by gold wires (width 2 μm, current 4.4 mA) and a homogeneous field $\boldsymbol{B}_b = (-0.1, -4.2, -1.6)$ Gauss. The color of the wire indicates the current density obtained from a finite element simulation. The Co coupling magnet on the tip of the Si mechanical oscillator is located directly below the center of the trapping potential. Compensation magnets on each side of the tip reduce the distortion of the trapping potential which is due to the static field of the coupling magnet. (b) Trapping potential in the yz-plane intersecting the resonator. The trap minimum is at $(y_0, z_0) = (1.5, 0.0)$ μm, the trap frequencies are $\omega_{x,y,z}/2\pi = (8.9, 9.7, 1.2)$ kHz for the $|1, -1\rangle$ state. The static field of the magnets causes a repulsive potential around $y = z = 0$. The attractive Casimir-Polder surface potential is visible for $y \to 0$. The orange area in the trap center shows the extension of the BEC.

$(7.0, 0.2, 0.1)$ μm, yielding an effective mass $m_{eff} = 3 \cdot 10^{-16}$ kg and a fundamental mode eigenfrequency $\omega_r/2\pi = 1.12$ MHz. It carries a single domain magnet with dimensions $(l_m, w_m, t_m) = (1.3, 0.2, 0.08)$ μm at its tip. This magnet is enclosed by two compensation magnets with the same cross section which have each a length of > 5 μm. The gap between the magnets is 200nm. The magnetic trapping potential is provided by an Ω-shaped wire and a straight wire of width 2 μm, which carry a current of 4.4 mA, and a homogeneous field $\boldsymbol{B}_b = (-0.1, -4.2, -1.6)$ Gauss. The center of the magnetic trapping potential is 1.5 μm above the magnet.

Atoms are lost out of the trap with the loss rate $\gamma = \gamma_{tbl} + \gamma_0$, where the rate of background gas collisions and atom-surface interactions γ_0 is dominated by the three-body loss rate $\gamma_{tbl} = 2.2 \times 10^{-12} s^{7/5} (N\omega_x\omega_y\omega_z)^{4/5}$ for N atoms. The magnetic trap is optimized to achieve maximum Γ_r for minimal background losses γ.

Fig. 5.7 (b) shows the magnetic trapping potential in the yz-plane intersecting the long axis of the mechanical oscillator. A BEC of ^{87}Rb atoms in the $|1, -1 >$ state is confined with trap frequencies $\omega_{x,y,z}/2\pi = (8.9, 9.7, 1.2)$ kHz (orange area in the trapping potential). The repulsive potential around $y = z = 0$ is due to the static field of the magnetic structures. The attractive Casimir-Polder surface potential

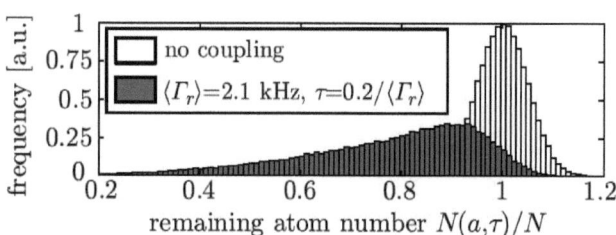

Figure 5.8: Coupling of a BEC to a thermally driven cantilever at $T = 300$ K. Simulated histogram of the fraction of atoms remaining in the trap after a time τ, including background losses. For comparison, the atom number distribution without coupling is shown. Fluctuations of 5 % in the atom number due to technical noise are assumed.

modifies the potential for $y \to 0$.

In a coupling experiment of a BEC to the mechanical oscillator at room temperature, the mechanical oscillator is in a thermal state. It performs oscillations with random phase ϕ and amplitude $a(t)$ which change on a time scale $\kappa^{-1} = (\omega_r/2Q)^{-1}$. In a single shot of the experiment, the mechanical oscillator is coupled to a BEC and drives spin flips to untrapped motional states during an interaction time $\tau \ll \kappa^{-1}$. The number of remaining atoms $N(a,\tau) = N\exp[-\Gamma_r(a)\tau]$ is measured with absorption imaging. For subsequent shots of the experiment, one expects a fluctuating $N(a,\tau)$ due to fluctuating $a(t)$. Fig. 5.8 shows the simulated signal $N(a,\tau)/N$ in a histogram. The histogram reflects the exponential distribution of the thermal phonon number in the mechanical oscillator as $\Gamma_r \propto a^2 \propto n_{th} = [\exp(\hbar\omega_r/k_BT) - 1)]^{-1}$.

For a BEC of $N = 10^3$, one calculates $\langle\Gamma_r\rangle = 2.1$ kHz for $r_c = 1/\sqrt{3}$. With a mechanical quality factor $Q = 5 \times 10^3$, and an interaction time $\tau = 0.2/\langle\Gamma_r\rangle$, we get $\kappa\tau = 0.07$. The background losses γ are almost negligible in comparison to the rate Γ_r, at which atoms are coupled out of the BEC, $\gamma = 0.01\langle\Gamma_r\rangle$. This parameter estimate shows that this chip layout could allow to resolve the thermal motion of the mechanical oscillator experimentally via observing the trap loss from a BEC.

5.4 Mechanical cavity QED

In a room temperature experiment, the mechanical oscillator discussed above has a thermal occupation $n_{th} \approx 5 \times 10^6$ which exceeds the number of trapped atoms

5.4 Mechanical cavity QED

$N = 1000$, i.e. the mechanical oscillators' thermal state is almost not influenced by the coupling. This changes at low temperatures, where $n_{th} \approx N$. Coherent energy exchange occurs like this: if an atom changes its state, the number of thermal phonons in the resonator changes by one. Depending on the initial state of the BEC, the phonon number in the mechanical oscillator can increase or decrease. The condition $n_{th} \approx N$ is fulfilled for $N = 1 \times 10^4$ and a bath temperature of $T = 0.5$ K.

In the following, we describe the quantum dynamics of the coupled system in the spirit of cQED. We consider transitions between two trapped atomic states $|0\rangle$ and $|1\rangle$. In a magnetic trap, the atoms in different hyperfine states experience different trapping potentials and trap frequencies. For the sake of simplicity, we consider optically trapped atoms which provides identical traps for all hyperfine states, and which could be $|0\rangle \equiv |2,1\rangle$ and $|1\rangle \equiv |2,2\rangle$ in an experimental realization. The transition $|0\rangle \leftrightarrow |1\rangle$ can be isolated from other hyperfine levels m_F by using e.g. microwave radiation to introduce m_F-dependent energy level shifts, or by making use of the quadratic Zeeman effect.

In this situation, N two-level atoms with level spacing $\hbar\omega_L$ can be described by a collective spin $S = N/2$ with hamiltonian $H_{BEC} = \hbar\omega_L \hat{S}_z$ and eigenstates $|S, m_s\rangle$, $|m_s| \leq S$. The Hamiltonian of the quantized mechanical oscillator is $H_r = \hbar\omega_r a^\dagger a$, where $a^\dagger$ (a) is the creation (annihilation) operator for phonons in the fundamental mechanical mode. The coupling Hamiltonian is obtained by replacing $\sqrt{2}g_F F_x \to \hat{S}_x$ and $a(t) \to a_{qm}(\hat{a}^\dagger + \hat{a})$, with the rms amplitude of the quantum mechanical zero point motion $a_{qm} = \sqrt{\hbar/2m_{eff}\omega_r}$. The coupled system is described by the Hamiltonian $H = H_r + H_{BEC} + H_Z$. With $\hat{S}^\pm = \hat{S}_x \pm i\hat{S}_y$ and the rotating wave approximation, we obtain the Tavis-Cummings Hamiltonian

$$H = \hbar\omega_r \hat{a}^\dagger \hat{a} + \hbar\omega_L \hat{S}_z + \hbar g(\hat{S}^+ \hat{a} + \hat{S}^- \hat{a}^\dagger), \quad (5.5)$$

with single atom – single phonon coupling constant $g = \mu_B G_m a_{qm}/\sqrt{8}\hbar$.

The situation is analogous to cavity QED experiments. In our case, the role of the electromagnetic field in an optical or a microwave cavity is taken over by the phonon field of a single mode of the mechanical oscillator. The single-atom single-phonon strong coupling regime is reached for $g > (\kappa, \gamma)$ in cavity QED. Due to the coupling to the thermal bath, one has to achieve $g > (\kappa n_{th}, \gamma) = (k_B T/2Q\hbar, \gamma)$ in the configuration considered here.

We show that the strong coupling regime for a single atom and a single mode of the mechanical oscillator can be reached for realistic parameters. The atom is trapped at a distance $y_0 = 250$nm with $\bar{\omega}_t/2\pi = 250$ kHz. The high trapping frequencies are possible due the absence of three-body collisions ($N = 1$). The mechanical oscillator consists of silicon, and has the dimensions $(l, w, t) = (8.0, 0.3, 0.05)$ μm. It carries a cobalt magnet with $(l_m, w_m, t_m) = (250, 50, 80)$nm and we assume a distance of the atom to the surface of the magnet of $d = 100$nm. The fundamental mode frequency is $\omega_r = 2.8$ MHz.

In order to reach the strong coupling regime, the mechanical oscillator has to have a high mechanical quality factor and a small bath temperature. Quality factors of $Q = 1.2 \times 10^5$ have been demonstrated [12] at $T = 100$mK with a mechanical oscillator which could be functionalized with a single domain magnet. Temperatures as low as $T = 100$ μK were reached and maintained in a nuclear demagnetization cryostat [133]. For these parameters, we obtain $(g\sqrt{N}, \kappa n_{th}, \gamma) = 2\pi \times (62, 55, 0.3)$ Hz with $N = 1$, which shows that the strong coupling regime could be reached for a single atom.

5.5 Steps towards the experimental realization

This chapter describes the fabrication process that was developed in order to realize the chip structure discussed in section 5.2. We fit 12 structures similar to the structure shown in Fig. 5.7 (a) onto the Mo chip, each containing a mechanical oscillator, cobalt magnets and gold wires for magnetic trapping. In addition to single-side clamped mechanical oscillators we also fabricate double-side clamped mechanical oscillators. Fig. 5.9 shows the layout of the Mo chip.

The fabrication process involves optical and electron beam lithography, evaporation of metallic material, as well as etching techniques [57]. In addition to the considerations in 5.2, the flow of the fabrication process is influenced by the following considerations.

- The mechanical oscillator is fabricated with a standard technique, which is based on electron beam lithography and etching techniques carried out on a Silicon-On-Insulator wafer (SOI wafer). The oscillator is patterned out of the top (monocrystalline) silicon layer, and left freestanding after removal of the sacrificial layer.

- The magnetic structure is defined with electron beam lithography and deposition of ferromagnetic cobalt. This results in a magnetic single domain structure.

- The thickness of the gold wires is determined by the currents of up to $I = 10$ mA which are guided through the wires in the central chip region in order to provide the magnetic trapping potential for the BEC. A safe value for the

5.5 Steps towards the experimental realization

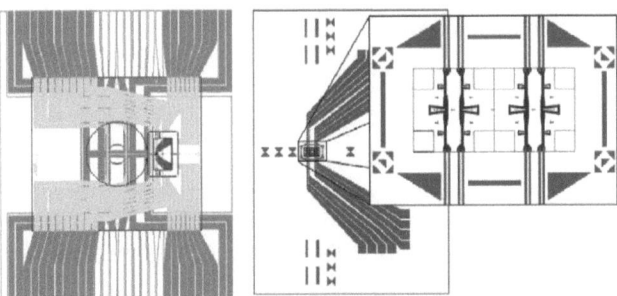

Figure 5.9: The atom chip consists of 3 chips which are glued on top of each other. The base chip closes the vacuum glass cell, and provides a feedthrough for the wires fabricated with optical lithography on the base chip, which provides a quadrupole potential for a magneto-optical trap (mirror-MOT). The smaller transport chip (light grey) provides magnetic fields for a first Ioffe-Pritchard trap and contains a long wire for creating a waveguide for transporting the atoms from the MOT region to the Mo chip (right hand side). The Mo chip carries one layer of bond pads and gold wires that guide currents to the central chip region, where 12 structures (similar) to the layout discussed in Fig. 5.7 (a) are fabricated.

maximum current density $j_{max} \approx 1 \times 10^{10}$ A/m^2 which a gold wire can support should not be exceeded. The smallest width of the gold wires is 2 μm, such that a thickness of the gold layer $\geq$ 500nm is favourable.

- The gold wires in the central region of the Mo chip are defined with electron beam lithography as they have to be fabricated subsequent to the magnetic structures with an alignment uncertainty < 1 μm. This can not be achieved with optical lithography.

- The gold wires on the Mo chip are electrically connected to the transport and the base chip with bond wire connections. Transport and base chip are similar to the chip package used for the experiments described in chapter 3. The Mo chip has to have bond pads for the electric connection and gold wires for guiding the currents to the central chip region. These structures are fabricated with optical lithography.

- The distance between the wires on the Mo chip and the wires on the transport chip should not exceed 100 μm in order to facilitate loading of the atoms into a magnetic trap which is provided by a current flow through wires in the central chip region of the Mo chip. As the layer of glue between the two chips has a typical thickness of 30 μm, the thickness of the Mo chip should be equal or less than 70 μm.

5.5.1 Fabrication methods

Lithography

We use lithography techniques to expose a resist coated chip in order to transfer the desired structure into the resist. Resists consist of polymer chains, which are either broken into smaller chains (positive resist) or crosslinked (negative resist) during the exposure process. After exposure, the resist is exposed to a developer which dissolves the resist: the shorter the length of a polymer chain, the higher the rate of dissolution. Choosing an appropriate development time, a positive (negative) resist can be dissolved completely in the exposed (unexposed) areas.

Fig. 5.10 shows the principle of the employed lithography techniques. In optical lithography the desired pattern is imprinted into an optically dense chromium layer on a glass substrate, such that a shadow mask is formed. The mask is positioned on top of the resist coated chip and shadows it partially according to the imprinted pattern. The pattern is transferred into the resist layer via exposure to UV light and subsequent dissolution of the resist. The resolution of optical lithography is limited by the wavelength of the UV light.

In comparison to optical lithography where the whole area is exposed in a parallel procedure, electron beam lithography exposes the resist at discrete grid points in a serial manner. The electron beam is positioned by applying electric potentials to capacitor plates arranged parallel to the electron beam path. To expose an area, the electron beam is unblocked for a set time. Due to backscattered electrons the neighbourhood of a grid point is exposed to a certain electron dose as well. This so called proximity effect transforms grid-like exposure into an exposed, connected area. In practical cases, the resolution of electron beam lithography is limited due to the proximity effect.

Material deposition

We deposit metallic material onto the resist pattern resulting from lithography and development of the resist. The material is heated with a focussed electron beam in a UHV vacuum chamber and evaporates. The chip is mounted at ≈ 30 cm distance from the material source. The evaporated metal adheres on bare substrate and on resist. After deposition of a material layer of the desired thickness, the chip is immersed into a solvent which removes the resist and the material evaporated on top, but leaves the material that was evaporated onto the bare substrate (lift-off).

5.5 Steps towards the experimental realization

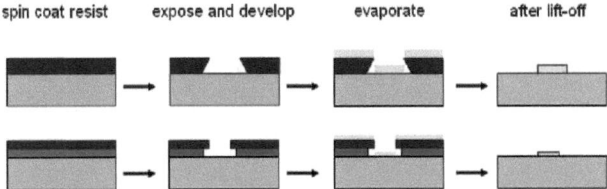

Figure 5.10: (left to right): A wafer is spin coated with a resist. The chip is selectively shadowed with a mask and exposed to UV light or scanned with a focused electron beam. Both methods change the relative solubility of exposed/not exposed resist to a developer. This defines a resist mask which partially covers the chip surface. Material is evaporated and coats the whole chip. The wafer is immersed into a solvent, which leaves only the material which is evaporated onto the bare substrate. The undercut (upper series) facilitates lift-off, and can also be achieved with a multilayer resist structure in e.g. electron beam lithography (lower series).

Etching techniques − freestanding structures

A Silicon-On-Insulator (SOI) wafer is a stack of three layers, a handle layer of silicon that supports a sacrificial oxide layer and a silicon layer on top. The top silicon layer is selectively removed in a reactive ion etching process, where e.g. a metallic etch mask is deposited on the chip surface. In this process, the chip is placed between two capacitor plates, which accelerate ionized molecules perpendicular to the chip surface, resulting in physical, anisotropic material ablation and chemical etching. The areas which should remain, are protected with an etch mask, which is removed after etching. The oxide that still supports the remained parts of the top silicon layer is removed partially with a chemical wet etch, which suspends the top silicon layer partially for a sufficiently long exposure to the etchant. Fig. 5.11 illustrates the process flow.

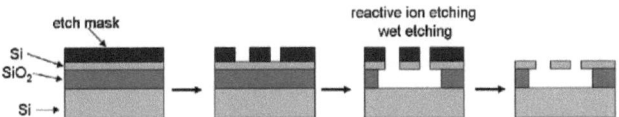

Figure 5.11: An SOI wafer is partially protected with an etch mask. Reactive ion etching is used to remove the top silicon layer anisotropically. An isotropic wet etch removes the oxide layer below the top silicon layer depending on the etching rate and time.

5.5.2 Process flow

The fabrication process is summarized in Fig. 5.12 and in the caption. The order of the process steps is determined by the design and fabrication methods. In particular, there is no possibility to exchange the order of any two of the steps. The main text of this section contains detailed information about the individual steps. Process parameters can be found in the appendix.

Substrate

We purchased Unibond SOI wafers at Soitec SA, France, where the Smart Cut Process (TM) is applied in wafer fabrication: a monocrystalline silicon wafer is cleaned and oxidized. Hydrogen ions are implanted into a set depth. The wafer is bonded to a monocrystalline silicon handle wafer, and cleaved at the surface that is defined and weakened by implanted hydrogen ions. SOI wafers fabricated with this process come with a thin and unstressed top silicon layer.

The monocrystalline top silicon layer of our wafer material has a thickness of 100nm. The silicon is p-type doped with boron, and has a maximum resistivity of 22.5Ω cm, which is high in comparison to evaporated gold that we use to guide currents on the chip. The orientation of the silicon monocrystal is $\langle 100 \rangle$. The sacrificial oxide layer and the handle layer have a thickness of 400nm and 675 μm, respectively. The diameter of the wafer is 150 mm.

To reduce the thickness of the the wafer to 70 μm, we first tried to grind the handle layer of the SOI wafer with machines available in our machine shop. However, as this proved to be not reproducible, we sent a wafer to the Fraunhofer IZM, which has equipment in place to reduce the thickness of wafers. The thickness of the handle layer was reduced to 46 μm ± 5 μm, and the thinned wafer was cut into pieces (chips) of 5 mm×7 mm size.

Optical lithography

A drawback in a one step optical lithography as described in 5.5.1 is deficient quality of the sidewalls of eventually deposited material. If an electric connection is to be established across conducting material deposited in several lithography steps, the electrical connection can be unsatisfactory depending on the slope of the sidewalls. This tends to be worse with increasing thickness of the deposited material layer.

Image reversal lithography offers good edge quality and the possibility to control the slope of the sidewalls, at the cost of a more complicated process (see Fig.5.13).

5.5 Steps towards the experimental realization

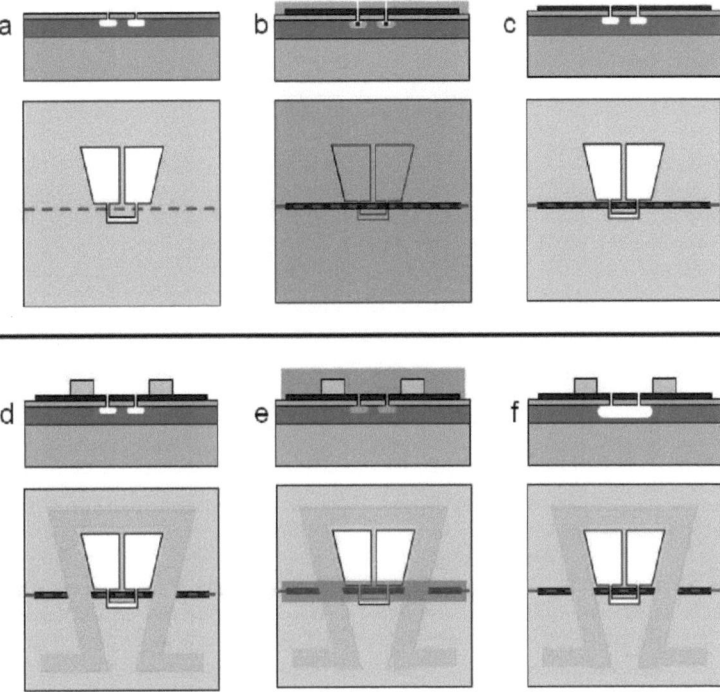

Figure 5.12: After patterning bond pads and wire structures with optical lithography, all further processing employs electron beam lithography. The schematics show the cross section (top) perpendicular to the chip surface along the dashed line in the top view (bottom). (a) Deposition of an etch mask for reactive ion etching with electron beam lithography and evaporation of a metal. The mechanical oscillator is shaped laterally and alignment markers for later lithography steps are provided. The top silicon layer is selectively removed with reactive ion etching, followed by an oxide wet etch that just underetches the rim of the top silicon layer without leaving structures freestanding. (b) Evaporation of a continuous stripe of cobalt across the mechanical oscillator. The topography resulting from the previous step separates the magnet on the mechanical oscillator from the enclosing magnets by a gap. Cobalt evaporated into the gap is removed in the subsequent liftoff, the result is shown in (c). (d) Deposition of gold wires for providing the trapping potential using electron beam lithography and evaporation of gold. (e) Enclosure of the magnetic structure in a matrix of exposed, negative resist for protection of cobalt against contact with hydrofluoric acid in the following step. (f) Wet etch with hydrofluoric acid to remove the supporting oxide below the mechanical oscillator, and subsequent removal of the resist matrix from the previous step.

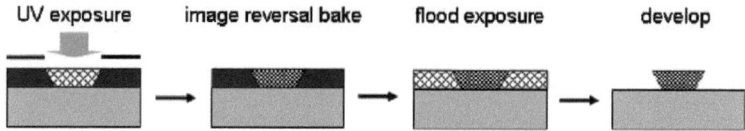

Figure 5.13: The chip substrate is spin coated with an image reversal resist. UV exposure cross-links the exposed resist and renders it insoluble in a developer. Flood exposure makes the resist shadowed in the first exposure step soluble to the developer, and allows to remove it in the final development step.

The Piranha[1]-cleaned chip substrate is spin coated with the image reversal resist AZ 5214 E[2] with a targeted thickness of the resist layer of 1.6 μm. After baking the chip on a hotplate, the resist is exposed for a first time. The first exposure with a chromium mask transfers the desired pattern into the resist, and has to be carried out carefully to achieve a very good resolution. Therefore the distance between mask and resist surface should be minimized. Resist residues close to the edges of a chip, that emerge during spin coating can be eventually removed with lens cleaning paper and acetone. In the image reversal bake, the resist polymers in the exposed region are cross-linked, and thus made insoluble towards the developer. The critical parameter of image reversal baking is the temperature which determines the slope of the resist sidewalls. Very reproducible results are obtained if the temperature within different runs varies by less than 0.1°C. The image reversal bake is followed by a flood exposure of the whole chip towards UV light without mask, which makes the resist shadowed in the first exposure step soluble to the developer.

Evaporation of gold and subsequent lift-off allow to fabricate gold wires with a thickness up to 1 μm with a resolution < 0.5 μm. The positions of structures fabricated in later steps with electron beam lithography are referenced to the structures in this gold layer.

Mechanical oscillator − Step (a)

Fabrication of magnetic structures and gold wires in the central chip region has to be done before the wet etch which suspends the oscillator, as spinning on a resist for lithography is not compatible with a fragile single side clamped oscillator. We split the fabrication of free standing structures into two parts.

We start by structuring the top silicon layer with reactive ion etching. We choose

---

[1] $H_2SO_4 : H_2O_2 = 4 : 1$
[2] Clariant

5.5 Steps towards the experimental realization 105

cobalt as a mask material, which can in principle be used as a mask material for reactive ion etching processes employing SF_6, $SiCl_4$ and CF_4. Already a relatively thin layer resists the etching procedure sufficiently long such that the top silicon layer can be completely removed without damaging the silicon in the protected areas. After the etch process, the etch mask is removed in a Piranha cleaning step.

In order to define the lateral shape, we employ one step of electron beam lithography and evaporate a cobalt etch mask. We also protect the whole area where the top silicon layer should not be removed in the etching process, e.g. at the areas where cobalt and gold are deposited in later steps. If the thickness of the evaporated layer is small in comparison to the resist layer, one can use a single resist layer consisting of long polymer chains, which results also in a better resolution. We use a single layer of PMMA 500000 for electron beam lithography, where we define the shape of the mechanical oscillators. After exposure and development, we evaporate 35nm of cobalt as an etch mask and use SF_6 or CF_4 as etchant in reactive ion etching. The etch mask is removed with piranha. In this process, we reproducibly achieve almost vertical sidewalls of the etched silicon layer and gap widths down to 100nm. Finally, we employ hydrofluoric acid to partially remove the sacrificial oxide layer, but without leaving the mechanical oscillators freestanding. Several steps of electron beam lithography and material deposition are carried out before the sacrificial layer below the mechanical oscillator is completely removed.

Note, that this is the last step in the process flow, where we can perform a careful cleaning of the chip. As soon as cobalt is deposited, cleaning of the chip with e.g. piranha is no longer possible.

Lithography markers In step (a) we also fabricate markers on the chip which play an important role in aligning structures fabricated in the individual steps relatively to each another. In each lithography step, the chip and the coordinate system of the electron beam lithography system have to be matched. This is done iteratively by assigning software coordinates to at least three markers located on the chip by taking scanning electron beam micrographs of these.

There are several methods to define markers with very high resolution in electron beam lithography. One could think of an individual step of electron beam lithography which is optimized to serve this purpose. Such procedures typically yield an alignment precision in subsequent steps of ±30nm. We fabricate marker structures in the same lithography step as the mechanical oscillators. This means, that the markers are not out of gold as usually, but structured out of the top silicon layer. As the imaging contrast of these markers, especially if they are below a layer of resist, is not very good on a scanning electron micrograph, we expose the area around the markers of the spin coated chip with the electron beam, and remove the exposed resist with developer, leaving the resist on the unexposed areas. The markers without resist coverage can be very well resolved and allow for precise alignment in the subsequent lithography step, where magnets or gold wires are defined in the remained resist.

Magnetic structures – Step (b) and Step (c)

We employ electron beam lithography on a multilayer PMMA resist system for defining the shape of the magnetic structures. The two lower layers (PMMA 150000) provide planarization for the top layer (PMMA 500000). Evaporation of cobalt goes along with an increased thermal load on sample and resist. The evaporation process differs from evaporation of gold in a higher melting point temperature of $1495°C$ in comparison to $1064°C$. We reduce the thermal load by mounting the cobalt target with minimal thermal contact in the UHV chamber which allows to melt and evaporate cobalt from only a small area on top of the target. This reduces the required electron beam currents significantly, and therefore reduces the thermal load on the resist and facilitates the lift-off. Furthermore, we insert a small, water cooled aperture to reduce heating of the sample holder. With these precautions it is easily possible to deposit cobalt and carry out the subsequent lift-off. We still encounter a problem with baked resist residues at the edges of the cobalt structures, which are not soluble in commonly used solvents for lift-off. The residues can make the mechanical oscillator inoperative, if they bridge the gap between the magnet on the mechanical oscillator and the enclosing magnets. Carefull optimization of the exposure parameters is necessary to reduce the amount of resist residues.

A continuous stripe of cobalt is evaporated across the mechanical oscillator. The topography resulting from the previous step separates the magnet on the mechanical oscillator from the enclosing magnets by a gap, that is well defined with reactive ion etching. Cobalt evaporated into the gap is removed in the subsequent lift-off. This technique allows to achieve the desired small gap and a rectangular shape of the magnets.

Wire structure for magnetic trapping – Step (d)

Usually, the thickness of a metallic layer deposited onto a resist structure patterned with electron beam lithography does not exceed 100nm. However, the gold wires should have a thickness of 500nm to stand the desired currents. Similar to the procedure in the last chapter, we use several layers of PMMA resist to provide planarization and a sufficient resist thickness allowing for evaporation and a subsequent

5.5 Steps towards the experimental realization   107

clean lift-off for a gold layer of desired thickness. The requirements to the resolution in this step are not very high in measures of what would be possible with electron beam lithography. This allows to optimize the resist and exposure parameters to yield a high reproducibility.

Resist shelter – Step (e)

The mechanical oscillator is finally suspended with a wet etch employing hydrofluoric acid to etch the sacrificial oxide layer. However, cobalt is dissolved almost completely by hydrofluoric acid. As we can not change to another magnetic material due to the requirements to the magnet or to another etchant we follow the strategy to protect cobalt against hydrofluoric acid with a shelter.

The requirements to the material of the shelter are quite demanding: it should be possible to pattern it with high resolution, it should be impermeable to hydrofluoric acid and it should be removable without destroying the fragile mechanical oscillator. We find such a material with the negative resist maN 2403[3], which can be patterned with electron beam or optical lithography at a resolution down to 50nm. We pattern a box of negative resist around the magnetic structure. The sacrificial layer is removed with hydrofluoric acid. After wet etching, the shelter is removed with either a solvent and a critical point dryer, or with oxygen plasma cleaning. The latter procedure is an elegant way to circumvent critical point drying, that can be applied to any fragile mechanical structure: before etching, the structure can be clamped at several points with exposure of the negative resist. After the wet etch, the chip can be dried with a nitrogen gun, and the freestanding structure can be released with plasma cleaning.

We found that the negative resist is impermeable to hydrofluoric acid. We performed one step of optical lithography to define the magnets on SOI chips, which have been only exposed to one step of optical lithography and gold deposition. After Cobalt evaporation and lift-off, we enclosed the cobalt structures with a resist shelter and exposed the chip to hydrofluoric acid. After removal of the resist, we observed neither with electron microscopy nor with magnetic force microscopy a change of shape, electron reflectivity or magnetic properties. Hence, the negative resist is a suitable material for a resist shelter.

---

[3]MicroResist

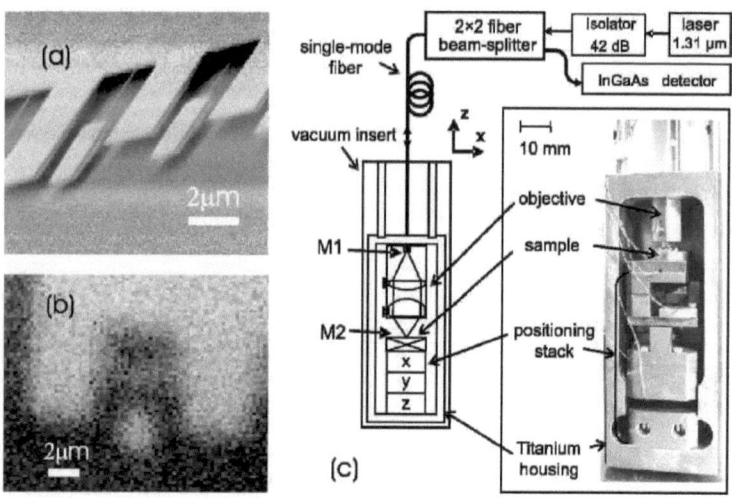

Figure 5.14: (a) Scanning electron micrograph of mechanical oscillators carrying a gold mirror for optical characterization, (b) Cavity *in situ* optical imaging of one mechanical oscillator, (c) Setup. Figure taken from [134]

5.5.3 Characterization

Mechanical oscillator

As it is difficult, to read out the motion of a sub-wavelength sized mechanical oscillators on the Mo chip, we fabricated similar, single-side clamped mechanical oscillators for a preliminary characterization. The oscillators are functionalized with a gold mirror for optical readout in an optical cavity, see [134]. We use essentially the same process which is employed to fabricate the mechanical oscillators on the Mo chip. We employ electron beam lithography to deposit a gold mirror. The lateral shape of the oscillators is transferred into the top silicon layer with reactive ion etching, employing an etch mask of negative resist maN 2403. After removal of the etch mask with a piranha etch, the structures are suspended in a wet etch with hydrofluoric acid. The chip is dried in a critical point dryer. The total length of the mechanical oscillator is 6 μm, the width is 200nm. At the tip, the Silicon forms a square of 2.4×2.4 μm^2. The thickness of the silicon layer is 100nm, and a gold layer of 90nm thickness is deposited to enhance the reflectivity. Fig. 5.14 (a) shows a scanning electron micrograph of the mechanical oscillator.

5.5 Steps towards the experimental realization

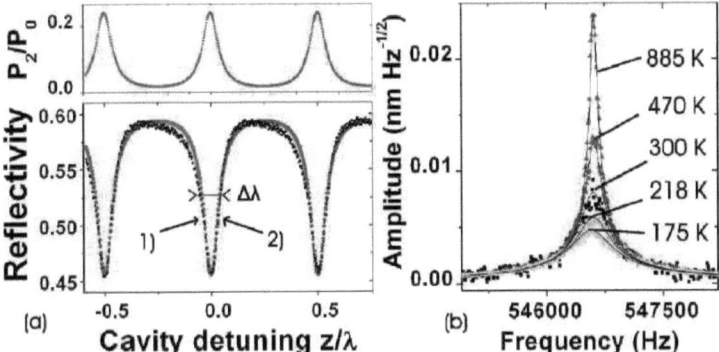

Figure 5.15: (a) Bottom: Measured (black) and modeled (red) reflectivity of the cavity. Top: Simulated ratio of the light power sent on the cavity formed by the mirrors M1 and M2. (b) Thermal motion amplitude spectra for negative cavity detuning. Figure taken from [134]

The setup is illustrated in Fig. 5.14 (c). A Fabry-Perot cavity is formed between the mirror on the mechanical oscillator (M2) and a gold coated end facet of a single mode glas fiber (M1). The reflected signal from the cavity is analyzed with a spectrum analyzer and allows one to extract a mechanical quality factor $Q \approx 1000$ and fundamental mode eigenfrequencies that match the calculated eigenfrequencies with a deviation smaller than a factor of 2. Fig. 5.14 (b) shows *in situ* imaging of the reflectivity of the substrate using the xy positioning stack to displace the chip.

In further experiments, the fundamental mode temperature of this mechanical oscillator was laser cooled to $T = 175$ K, employing a bolometric effect. Fig. 5.15 (a) shows the cavity reflected power versus detuning, and Fig. 5.15 (b) shows the cooling of the fundamental out of plane mode. Further details are given in [134].

Magnetic characterization

We use magnetic force microscopy (MFM) to characterize the domain structure of the fabricated cobalt magnets. We apply the procedure which is descibed in [132]. The measured quantity is the phase shift of the oscillating AFM cantilever due to the gradient of the force between the magnetic tip of the AFM cantilever and the magnetic structures on the sample. Fig. 5.16 shows a measurement of the compensation effect. The magnetic tip position is scanned in a plane which is

110   Coupling to the collective spin of a BEC

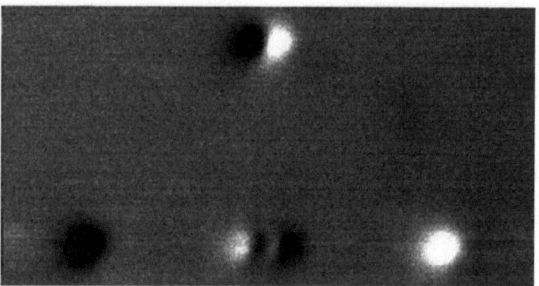 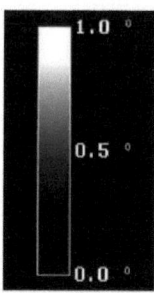

Figure 5.16: MFM measurement with a lift height of 700nm. (Top) Single magnet, (Bottom) single magnet of same dimensions, with two enclosing magnets. The static magnetic field gradient is attenuated due to the enclosing magnets by a factor of 15.

700nm above the surface. The measured signal at the top originates from a single domain magnet without enclosing magnets. The structure below has two additional enclosing magnets with a gap distance of $d = 150$nm, and the static magnetic field above the magnet in the center is obviously reduced. A measurement [132] shows that the magnetic field is attenuated due to the enclosing magnets by factor of 15.

5.5.4 Fabrication: Status

The fabrication process was optimized and the results of all steps satisfied the expectations – except the result of step (e), where the mechanical oscillator is suspended in a wet etch. Fig. 5.17 shows the chip structure before the wet etch is performed, Fig. 5.18 shows a prototype structure after wet etching. The surface of the top silicon layer is etched significantly during this step with hydrofluoric acid. As a secondary effect, the resist shelter which protects the deposited cobalt, was underetched such that hydrofluoric acid could attack and dissolve the cobalt structures.

These circumstances did not allow to fabricate the desired structures reproducibly. In all trials at least 80% of the cobalt structures were damaged, which was clearly visible in an electron beam micrograph. The magnetic properties of the magnets on the mechanical oscillator could not be characterized with magnetic force microscopy, because the the mechanical oscillators were destroyed from the AFM tip which is alternatingly operated in tapping and lift-mode in magnetic force microscopy.

In the first wet etch in step (a), we did not observe any damage of the top silicon layer. Hence, we attribute the surface degradation occuring in step (e) to a chemical reaction of contamination accumulated in the process steps later than step (a). After the deposition of the magnetic material in step (b), cleaning with Piranha is

5.5 Steps towards the experimental realization

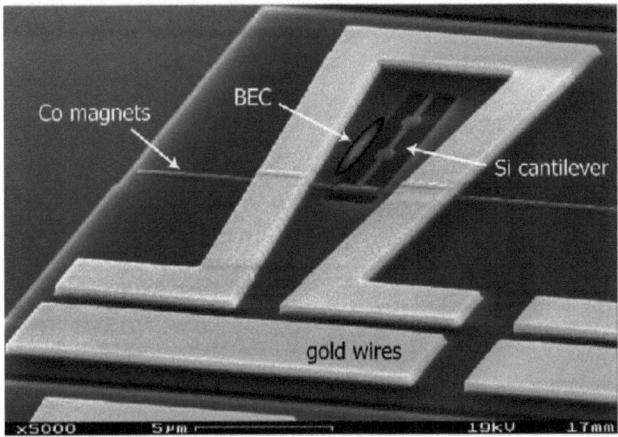

Figure 5.17: Scanning electron micrograph of the chip structure, before step (e) is performed. The BEC is sketched in red.

no longer possible since cobalt is dissolved in Piranha.

We analyze the source of the contamination. The chips were cleaned with piranha before the fabrication of cobalt magnets in step (b). In principle we have to suspect all substances used in later steps and any combinations of these:

- Substrate: silicon, silicon oxide
- Resists: PMMA and developer, negative resist and developer
- Solvents: water, isopropanol, acetone or ethanol
- Deposited material: gold, cobalt, titanium (adhesion promoter)
- Wet etchant: hydrofluoric acid and ammonium fluorid

The most suspicious candidate for contamination which interacts in the wet etch step with the etchant and silicon are the resists and the deposited materials. To exclude the resists and developers, we performed tests with virgin chips, coated them with the negative resist (or alternatively PMMA), exposed them partially and used the respective developer. Exposure of these chips to hydrofluoric acid did not result in degradation of the top silicon layer. This excludes unfavourable behaviour of the resist. In addition, we studied the behaviour of metal on the chip substrate during

Figure 5.18: Scanning electron micrograph of the chip structure, after step (e) is performed.

5.5 Steps towards the experimental realization 113

the wet etching. We evaporated a thin gold layer onto a silicon chip. We exposed the chip to the etchant consisting of hydrofluoric acid and ammonium fluorid, and observed a degradation of the silicon surface. With a similar test, we saw that cobalt in combination with hydrofluoric acid by contrast does not lead to significant surface degradation.

We identified one major mechanism responsible for degradation, which involves predominantly the interplay of gold and the wet etchant. Ideally, the metal which is deposited during evaporation should be removed from the chip surface with the underlying resist. However, after lift-off traces of gold may remain on the chip surface, especially if the lift-off is done as gentle as possible in order not to damage the structures. During the wet etch with hydrofluoric acid, traces of metal on silicon can modify the etch characteristics of silicon. At the interface of silicon and gold a contact potential is formed, and leads to a local electric field, which influences the etching process. Recently, metal assisted etching was studied in several groups [135, 136, 137, 138]. This etching technique makes use of a thin metal layer deposited on a silicon wafer. Etching with hydrofluoric acid and an oxidizing agent allows to etch silicon which is in contact with the metal. Depending on material, oxidizing agent and doping, the etching process can be controlled, and e.g. nano-sized pillar fields can be fabricated.

From this analysis, we conclude that contamination of the silicon surface with traces of gold or titanium may be responsible for the degradation of the top silicon layer, and the insufficient adhesion of the resist shelter.

5.5.5 Fabrication: Perspective

The philosophy of the fabrication so far aimed at integrating all structures needed for the experiment on one chip. In practice, this resulted in a complicated process flow with many critical steps. Fabrication is very challenging as waste has to be almost avoided. Even if a usable chip is fabricated, it would not be possible to check the magnetic properties before integrating it into the experiment. Furthermore, it would be very difficult to read out the oscillation of the mechanical oscillator independently with e.g. an optical technique, as it is not functionalized with a mirror. These uncertainties and the fact that it takes 3 to 6 months to integrate a chip into an atom chip cell and to achieve Bose-Einstein condensation, raise the demands towards the perfection of a chip, that might be promising to integrate into an atom chip experiment, to a very high level. Degradation of the top silicon layer could be circumvented with a change of the substrate. As silicon nitride is chemically more resistive, the process could be adapted to a wafer similar to an SOI wafer, but with a top layer out of silicon nitride instead of silicon. This would reduce the degradation of the silicon surface occuring in step (e). First trials showed that an exchange of

Coupling to the collective spin of a BEC

the substrate would solve this problem.

In the following, I describe an alternative process which is easier, but requires a reactive ion etching process that was not available in the cleanroom facilities used for the fabrication in this thesis. The idea is to fabricate mechanical oscillators at the edge of a chip, to make use of the anisotropic characteristics of cobalt evaporation and to mount the chip close to wires on an atom chip providing a magnetic potential, or to use a focused laser beam to provide an optical potential. Mechanical oscillators fabricated at a chip edge are employed by other groups [139] for measuring magnetic forces with high sensitivity [5]. The oscillators used there are optimized to have small spring constants. The oscillators are functionalized with a magnet which is evaporated onto the flat top surface of the oscillators.

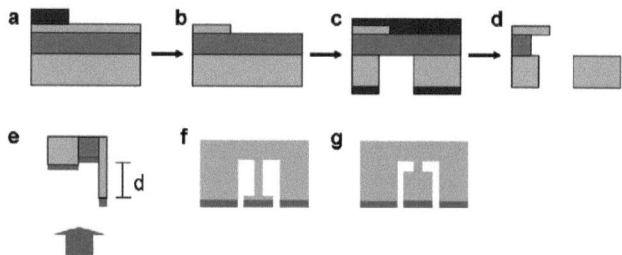

Figure 5.19: Process for fabrication of a mechanical oscillator at a chip edge. (a) Pattern lateral dimensions of the mechanical oscillator, fabrication of an etch mask. b) Reactive ion etching and removal of the mask. (c) Pattern etch mask for handle layer etch, remove handle layer selectively with deep reactive ion etching. Protect structures on top of the chip with resist. (d) Remove resists with a solvent, remove oxide with hydrofluoric acid. (e) Cobalt can be deposited without the need for lithography techniques, simply by orienting the structure carefully during cobalt evaporation. The distance d can be chosen such that the impact of magnetic cobalt evaporated onto the support is minimized. (f) and (g) show a top view of mechanical oscillators fabricated with this procedure. The oscillator shown in (g) provides a surface which could be used for optical readout of the motion.

The fabrication process involves the removal of the handle layer of an SOI wafer with reactive ion etching. The process flow is described in the caption of Fig. 5.19. With this process, monocrystalline cantilevers with dimensions up to $(l, w, t) = (120, 3, 0.1)$ μm can be fabricated [140]. One could employ this process flow to fabricate single side clamped mechanical oscillators, sticking out of a chip as shown in Fig. 5.19. In order to deposit the magnetic structure, the mechanical oscillator is oriented in a different way in comparison to the work cited above. Here, the long

5.5 Steps towards the experimental realization

axis of the mechanical oscillator is oriented parallel to the beam of evaporated cobalt in the evaporation chamber, such that cobalt sticks only to the small edge of the mechanical oscillator. Using a small aperture, one can make sure that the cobalt 'beam' is approximately collimated. This procedure allows for the deposition of a single domain magnet on the mechanical oscillator and enclosing magnets nearby.

This relatively easy process has the advantage, that deposition of cobalt is the last process step, i.e. it is unlikely that it looses its ferromagnetic properties in further process steps. A square at the tip of the mechanical oscillator could be used for optical readout, using the bare silicon as a partially reflective mirror. Furthermore it would be possible to fabricate a large number of similar structures along the edge, and to shift the magnetic trapping potential to a specific oscillator with dimple trap wires.

6 Outlook

The experiments in this thesis demonstrate the observation of interaction between microstructured mechanical oscillators and ultracold neutral atoms. In this chapter, I review the experimental achievements and sketch the prospectives of coupling mechanical oscillators to neutral atoms.

Coupling via the surface potential We demonstrated the coupling of a mechanical oscillator to a magnetically trapped BEC on an atom chip via the surface potential which arises from the oscillators surface. A nice feature of this coupling mechanism is that no additional functionalization of the mechanical oscillator is required. Thus, this coupling mechanism is also applicable to molecular scale oscillators which cannot be functionalized with a magnet or a mirror. One candidate for further experiments are suspended carbon nano tubes (CNTs), which can have high mechanical quality factors [141]. A single-wall CNT with a suitable eigenfrequency of $\omega/2\pi = 20$ kHz has a mass of $M = 2 \times 10^{-17}$ g [142], which is several orders of magnitude smaller than the mass $M = 5 \times 10^{-9}$ g of the mechanical oscillator used in the experiment reported in this thesis. Coupling of such a CNT to a BEC of $N = 1000$ atoms would achieve a mass difference of only two orders of magnitude. Hence, the coupling would be significantly stronger for a molecular scale oscillator from this perspective. On the other hand, the surface potential of a CNT is expected to be weaker by roughly one order of magnitude [143]. Currently, the coupling of static CNTs to ultracold atoms via the surface potential is investigated experimentally in the group of Prof. Joszef Fortagh at the Universität Tübingen.

Coupling via a 1D optical lattice We coupled a mechanical oscillator and thermal atoms via a 1D optical lattice, and observed the backaction of the atomic ensemble onto the oscillator. We compare the strength of the effect to theoretical predictions, and find reasonable agreement. This allows to make trustworthy theoretical estimates on how to improve the system. To enhance the coupling strength between ultracold atoms and mechanical oscillators in a motion-to-motion coupling as in the schemes discussed above, equal mass of two coupling partners is favourable. This results in an optimization such that the mass of the mechanical oscillator is reduced while the number of trapped atoms is increased. A first step in our experiment is to increase the number of trapped atoms. We load $N \approx 2 \times 10^6$ atoms from a mirror-MOT into the optical lattice, which is not very good in terms of atom number and temperature. Initially, our experimental setup was not designed to trap a large number of atoms and to perform Raman sideband cooling. This could be improved in a new setup, where one can apply Raman side-band cooling to prepare up to $N = 3 \times 10^8$ atoms in the ground state of a blue detuned 3D optical lattice, which was already demonstrated in [99]. In addition, the mass of the mechanical oscillator should be reduced. One possible way, which is currently investigated in our group, could be structuring of SiN membranes similar to the ones employed in our experiment. A focused ion beam is used to pattern double side clamped mechanical oscillators with a sufficient width which allows to reflect the lattice laser beam. Apart from these efforts, an enhanced membrane reflectivity[1] could increase the coupling strength directly due to the linear scaling with the power reflectivity R^2. The estimates in section 4.1.4 show, that improvements could allow for sympathetic ground state cooling of a cryogenically pre-cooled mechanical oscillator.

The limitation of the coupling strength due to the dependence on the square root of the mass ratio could be lifted, if the action of the lighter coupling partner onto the heavier one could be enhanced using some kind of 'lever'. In the coupling scheme via a 1D optical lattice, the action of the atoms onto the membrane is due to a dispersive interaction of the atom with the light field, and results in a power modulation of the running wave which travels towards the mechanical oscillator. The power modulation of the light field could be enhanced in an optical cavity, where already a single atom can modulate the light field considerably [144]. The coupling to the mechanical oscillator would be mediated by the light field in the optical cavity, similar as in the experiment described in [58], where the intracavity light field is used to cool the fundamental mode of a membrane oscillator .

A coupling scheme involving these ingredients was recently proposed in [38]. In this scheme, both the mechanical oscillator and a single, neutral atom are coupled to the cavity light field as shown in Fig. 6.1 (a). The modulation of the laser intensity results from the interaction of the light field with a membrane or an atom, and establishes the coupling. (b) The optical cavity modes are driven by laser light of two wavelengths, one blue and one red detuned with respect to the cavity response. (c) In addition, both laser frequencies are red detuned with respect to optical transitions, e.g. the $D_{1,2}$ line of an alkali atom, such that light of both laser wavelengths provide an attractive potential for the atom. (d) (left side) The atom is trapped in the effective potential provided by the superposition of the red-detuned laser fields (light and dark grey curves). (right side) The membrane thickness is small in comparison to the wavelength, which allows to position the membrane on the slope of

---

[1] However, an increase in the reflectivity is not straightforward to realize. A metallic coating increases the absorption of optical power significantly, whereas a dielectric coating increases the mass of the membrane oscillator. Moreover, it is likely that the mechanical quality factor decreases if material is deposited on the membrane.

the intensity profile. (e) The cavity resonances are shifted as indicated by the arrows in (b) due to the interaction of the membrane with the light field. As a consequence, the intensity of the red (blue) laser field decreases (increases), and goes along with a shift of the trap minimum. In this scheme for coupling motion-to-motion, the square root of the mass ratio does not enter into the coupling constant, due to the boundary conditions introduced by the mirrors. The interaction of the atom and the membrane does not rely on a direct interaction, as e.g. in the coupling of the center of mass modes in the membrane and the atomic ensemble via an optical lattice.

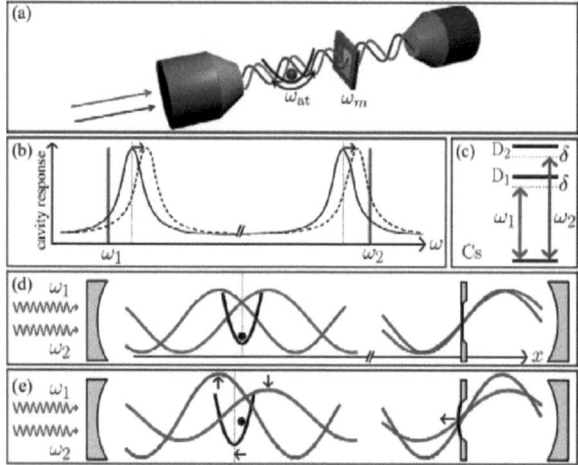

Figure 6.1: Strong coupling of the motion of a single atom and the motion of a SiN membrane via the intra-cavity light field. See text for the description. Figure taken from [38].

The exciting feature of this coupling scheme is that strong coupling of a single atom and a single mode of the mechanical oscillator seems to be feasible for realistic parameters and available technology. The experimental challenge is to operate a high finesse optical cavity at cryogenic temperatures, to achieve strong coupling of the light field to a single atom and to integrate a mechanical oscillator stably into the small gap between the cavity mirrors, which are a few tens of microns apart.

Coupling to the atomic spin The magnetic coupling of a mechanical oscillator to the spin of trapped atoms was investigated theoretically. The coupling relies on the magnetic field gradient field created by the magnet at the oscillators tip.

The theoretical formulation suggests that the coupling strength can be increased simply by increasing the gradient of the magnetic field, what could be achieved by enlarging the magnet. However, the magnetic field of the magnet strongly distorts the trapping potential, and it turned out in the analysis that the coupling strength is limited by the requirement, that the trapping potential for the BEC should not vanish. The impact of the magnet onto the trapping potential can be reduced with additional magnets, which enclose the magnet sitting on the mechanical oscillator. We demonstrated the resulting attenuation of the magnetic field experimentally. This effect should allow to reach a sufficiently high coupling strength for resolving the thermal motion of a nano-sized mechanical oscillator spectrally. A couple of years ago, our motivation was to investigate a suitable coupling mechanism between a mechanical oscillator and ultracold atoms, and to read out the thermal motion of a nano-sized mechanical oscillator. Spectrally resolved read out and opto-mechanical cooling [134] have been realized for the mechanical oscillator with a gold mirror which was presented in section 5.5.3 of this thesis. Based on our fabrication results, we propose a simplified procedure to functionalize nano-sized mechanical oscillators with single domain magnets. This could be used for an optimized version of this experimental design, or for fabrication of a mechanical oscillator which could be useful in a similar experiment. We started a collaboration on fabrication of mechanical oscillators with the group of Prof. Mikhail Lukin at Harvard University. This group is currently setting up an experiment which aims at coupling the mechanical motion of a nano-sized oscillator with a magnetic tip to single NV centers [145]. This experiment follows a scheme which is similar to the one that we have investigated [31], with a single NV center taking over the place of the BEC.

Appendix: Parameters of chip fabrication

Equipment Standard cleanroom facilities like e.g. hotplates are used; special equipment is listed below.

- **AFM** Atomic force microscope, Digital Instruments Dimension 3000 AFM.
- **UHV evaporation** material deposition system comprising a ultra-high vacuum chamber and an electron gun for evaporation.
- **Mask aligner** Karl Suss MJB3, with Hg UV-light source.
- **Plasma cleaner** providing oxygen plasma, Lab-Ash 100.
- **SEM** Scanning electron microscope, LEO 982 Digital Scanning Microscope equipped with Elphy for electron beam lithography.
- **Spin coater** Convac 1001S.
- **Ultrasound cleaner** Bandelin Sonorex Super.

Chemicals

- Standard chemicals like e.g. **Acetone**, **IPA** (Isopropanol), **MIBK**, **DI** (deionized) water, or acids and bases employed are VLSI selectipur quality throughout the process.
- **Piranha etch** mixture of sulphuric acid and hydrogen peroxide in volume ratio 3:1. It is seriously recommended to strictly avoid mixture of piranha with organic solvents, this results in an exothermal reaction.
- **HF** mixture of 1 part hydrofluoric acid and 7 parts ammoniumfluoride.
- **Plasma cleaning** in oxygen plasma cleaner.
- **US bath** place sample immersed in a solvent in the ultrasound cleaner.
- **Blow dry** with nitrogen gun.

Process parameters

In the following, the process steps for chip fabrication are described in detail. It is assumed, that the reader is familiar with cleanroom processing.

Cleaning of the substrate Cleaning of the substrate described in section 5.5.2. Piranha etch. Carefull rinse with DI in two separate beakers. Rinse with IPA. Blow dry.

Optical lithography Bake substrate for 5 min at 170°C on hotplate. Pause of 5 min between each of the following steps. Spin coat chip with AZ5214 E, 1s at 800 rpm, 30s at 3000 rpm. Bake for 50 s at 110°C on hotplate. Wipe off resist residues close to the edges of the chip with aceton. UV exposure with chromium mask for 3.5 s. Bake for 2 min at 120°C on hotplate. UV flood exposure for 40 s. Develop for 15 s to 30 s (depending on the age of the resist) with developer AZ 351:DI=1:4. Rinse with DI in two separate beakers, shake in first beaker for 1 min, leave for 5 min in second beaker. Bake on hotplate for 1.5 min at 100°C. Plasma cleaning for 10 s at 40 W and 2 Torr.

gold evaporation Clamp chip onto holder. Do not glue with UV compatible tape, as thin chips are too fragile to be removed applying force. Evaporate 3nm titaniumiumium and subsequently 400nm gold at 2.5 A/ s. Lift-off is no problem if the resist edges have an undercut. For Lift-off, immerse chip into Acetone for 20 min. Rinse with IPA. Blow dry.

Metal etch mask for reactive ion etching Bake substrate for 5 min at 170°C on hotplate. Spin coat chip with PMMA 500k 4%, 1s at 800 rpm, 30 s at 5000 rpm. Bake for 2 min at 170°C on hotplate. Expose resist in SEM at $U = 10$ kV with a dose $D_0 = 64$ $\mu C/cm^2$. Develop in MIBK:IPA=3:1 for 50 s. Rinse with IPA for 1min. Plasma cleaning for 10 s at 40 W and 2 Torr.

Alternative process: Resist etch mask for reactive ion etching Bake substrate for 5 min at 170°C on hotplate. Spin coat chip with maN 2403, 3 s at 800 rpm, 30 s at 3000 rpm. Bake chip for 60 s at 90°C on hotplate. Expose resist in SEM at $U = 20$ kV with a dose $D_0 = 50$ $\mu C/cm^2$. Develop in maD 532 for 70 s. Note: new developer should be applied for *each* chip. Rinse with DI for 1 min. Place for $t > 4$ min in a second beaker with DI. Blow dry. Plasma cleaning for 10 s at 40 W and 2 Torr.

Reactive ion etching Place chips in RIE-PP (homebuilt ion etching facility). Use a drop of Difelin oil to fix to the lower capacitor plate. Etch for 25 min at 100 W with a flow of 50 sccm CF_4 and a pressure $p = 5.8 \times 10^{-2}$ mBar.

Partial underetching Expose chip to HF for 50s. Note: use teflon tweezer instead of steel or carbon tweezer. Threefold rinse with DI. Piranha etch. Threefold rinse with DI. Blow dry.

Fabrication of cobalt magnets Bake substrate for 5 min at 170°C on hotplate. Spin coat chip with 2 layers PMMA 150K 4% and 1 layer PMMA 500k 4%, 1s at 800 rpm, 30 s at 5000 rpm. Bake for 2 min at 170°C on hotplate. Remove resist on top of the markers in order to achieve higher resolution in a first step of electron beam lithography at $U = 10$ kV with a dose $D_0 = 100$ μC/cm^2. Develop for 50 s in MIBK:IPA=3:1. Rinse for 1 min in IPA. Write magnet structures in a second step of electron beam lithography at $U = 10$ kV with a dose $D_0 = 175$ μC/cm^2. Develop for 60s in MIBK:IPA=3:1. Rinse for 1 min in IPA. Bake on hotplate for 1 min at 100°C. Plasma cleaning for 7 s at 40 W and 2 Torr. Evaporate 75 nm of cobalt at 0.3 A/ s. Lift-off with acetone. Rinse in IPA for 1 min. Immerse 10 min in Panasolve 180 at 50°C. Note, that the size of the written structure might differ from the fabricated structure due to proximity effect.

Fabrication of gold structures with electron beam lithography Bake substrate for 5 min at 170°C on hotplate. Spin coat chip with 2 layers PMMA 150k 16%, 1 s at 800 rpm, 30s at 3000 rpm. Spin coat chip with 2 layers PMMA 500k 16%, 1 s at 800 rpm, 30 s at 3000 rpm. Bake for 2 min at 170°C on hotplate. Remove resist on top of the markers in order to achieve higher resolution in a first step of electron beam lithography at $U = 10$ kV with a dose $D_0 = 100$ μC/cm^2. Develop for 50 s in MIBK:IPA=3:1. Rinse for 1 min in IPA. Write magnet structures in a second step of electron beam lithography at $U = 10$ kV with a dose $D_0 = 150$ μC/cm^2. Develop for 50 s in MIBK:IPA=3:1. Rinse for 1 min in IPA. Clamp chip onto holder for insertion into UHV evaporation chamber. Evaporate 3 nm titanium, and subsequently 400 nm gold at 2.5 A/ s. Lift-off is no problem due to an undercut of the resist edges. For Lift-off, immerse chip into Acetone for 20 min. Rinse with IPA. Blow dry.

Fabrication of resist shelter Bake substrate for 5 min at 170°C on hotplate. Spin coat chip with maN 2403, 3 s at 800 rpm, 30 s at 3000 rpm. Bake chip for 60 s at 90°C on hotplate. Expose resist in SEM at $U = 20$ kV with a dose $D_0 = 150$ μC/cm^2. Develop in maD 532 for 70 s. Note: new developer should be applied for *each* chip. Rinse with DI for 1 min. Place for $t > 4$ min in a second beaker with DI. Blow dry. Plasma cleaning for 10 s at 40 W and 2 Torr. Alternatively, development can be done with exposure to 1.6 % NaOH for 5 s.

Full suspension of the mechanical oscillator and removal of the resist shelter
Note, that this step did not work as expected. The following was tried, and showed the relatively best results. Exposure of chip to HF for 2 min. Threefold rinse with DI. Blow dry. Plasma cleaning for 500 s at 40 W and 2 Torr in order to remove the resist shelter completely. The maximum thickness of the shelter is 300 nm, the rate at which the resist is removed in the plasma cleaning process is > 1 nm/ s.

Bibliography

[1] K. L. Ekinci, Y. T. Yang, and M. L. Roukes, *Ultimate limits to inertial mass sensing based upon nanoelectromechanical systems*, J. Appl. Phys **95**, 2682 (2004).

[2] H. J. Mamin and D. Rugar, *Sub-attonewton force detection at millikelvin temperatures*, Appl. Phys. Lett. **79**, 3358 (2001).

[3] W. C. Fon, K. C. Schwab, J. M. Worlock, and M. L. Roukes, *Nanoscale and Phonon-Coupled Calorimetry with Sub-Attojoule/Kelvin Resolution*, Nano Letters **5**, 1968 (2005).

[4] K. Jensen, K. Kim, and A. Zettl, *An atomic-resolution nanomechanical mass sensor*, Nature Nanotechnology **3**, 533 (2008).

[5] D. Rugar, R. Budakian, H. J. Mamin, and B. W. Chui, *Single spin detection by magnetic resonance force microscopy*, Nature **430**, 329 (2004).

[6] T. J. Kippenberg and K. Vahala, *Cavity Optomechanics*, Opt. Exp. **15**, 17172 (2007).

[7] T. J. Kippenberg and K. Vahala, *Cavity Optomechanics: Back-Action at the Mesoscale*, Science **321**, 1172 (2008).

[8] F. Marquardt and S. M. Girvin, *Optomechanics*, Physics **2**, 40 (2009).

[9] I. Favero and K. Karrai, *Optomechanics of deformable optical cavities*, Nature Photon. **3**, 201 (2009).

[10] R. G. Knobel and A. N. Cleland, *Nanometre-scale displacement sensing using a single electron transistor*, Nature **424**, 291 (2003).

[11] M. D. LaHaye, O. Buu, B. Camarota, and K. C. Schwab, *Approaching the Quantum limit of a Nanomechanical Resonator*, Science **304**, 74 (2004).

[12] A. Naik, O. Buu, M. D. LaHaye, A. D. Armour, A. A. Clerk, M. P. Blencowe, , and K. C. Schwab, *Cooling a nanomechanical resonator with quantum back-action*, Nature **443**, 193 (2006).

[13] C. A. Regal, J. D. Teufel, and K. W. Lehnert, *Measuring nanomechanical motion with a microwave cavity interferometer*, Nature Phys. **5**, 555 (2008).

[14] A. D. O'Connel, M. Hofheinz, M. Ansmann, R. C. Bialczak, M. Lenander, E. Lucero, M. Neeley, D. Sank, H. Wang, M. Weides, J. Wenner, J. M. Martinis, and A. N. Cleland, *Quantum ground state and single-phonon control of a mechanical resonator*, Nature **464**, 697 (2010).

Bibliography

[15] T. W. Hänsch and A. L. Schawlow, *Cooling of gases by laser radiation*, Opt. Commun. **13**, 6869 (1975).

[16] The Nobel Prize in Physics 1997 and awarded to Steven Chu and Claude Cohen-Tannoudji and William D. Phillips .

[17] K. Jähne, C. Genes, K. Hammerer, M. Wallquist, E. S. Polzik, and P. Zoller, *Cavity-assisted squeezing of a mechanical oscillator*, Phys. Rev. A **79**, 063819 (2009).

[18] I. Bloch, *Ultracold quantum gases in optical lattices*, Nat. Phys. **1**, 23 (2005).

[19] M. W. Zwierlein, J. R. Abo-Shaeer, A. Schirotzek, C. H. Schunck, and W. Ketterle, *Vortices and superfluidity in a strongly interacting Fermi gas*, Nature **435**, 1047 (2005).

[20] S. Chu, *Cold atoms and quantum control*, Nature **416**, 206 (2002).

[21] M. F. Riedel, P. Böhi, Y. Li, T. W. Hänsch, A. Sinatra, and P. Treutlein, *Atom-chip-based generation of entanglement for quantum metrology*, Nature **464**, 1170 (2010).

[22] H. Häffner, W. Hänsel, C. F. Roos, J. Benhelm, D. Calkar, M. Chwalla, T. Körber, U. D. Rapol, M. Riebe, P. O. Schmidt, C. Becher, O. Gühne, W. Dür, and R. Blatt, *Scalable multiparticle entanglement of trapped ions*, Nature **438**, 643 (2005).

[23] M. Steffen, M. Ansmann, R. C. Bialczak, N. Katz, E. Lucero, R. Mc-Dermott, M. Neeley, E. M. Weig, A. N. Cleland, and J. M. Martinis, *Measurement of the Entanglement of Two Superconducting Qubits via State Tomography*, Science **313**, 1423 (2006).

[24] J. Majer, J. M. Chow, J. M. Gambetta, J. Koch, B. R. Johnson, J. A. Schreier, L. Frunzio, D. I. Schuster, A. A. Houck, A. Wallraff, A. Blais, M. H. Devoret, S. M. Girvin, and R. J. Schoelkopf, *Coupling superconducting qubits via a cavity bus*, Nature **449**, 443 (2007).

[25] E. Jaynes and F. Cummings, *Comparison of quantum and semiclassical radiation theories with application to the beam maser*, Proc. IEEE **51**, 89–109 (1963).

[26] C. Deutsch, F. Ramirez-Martinez, C. Lacroûte, F. Reinhard, T. Schneider, J. N. Fuchs, F. Piéchon, F. Laloë, J. Reichel, and P. Rosenbusch, *Spin Self-Rephasing and Very Long Coherence Times in a Trapped Atomic Ensemble*, Phys. Rev. Lett. **105**, 020401 (2010).

[27] D. J. Wineland, C. Monroe, W. M. Itano, D. Leibfried, B. E. King, and D. M.Meekhof, *Experimental Issues in Coherent Quantum-State Manipulation of Trapped Atomic Ions*, J. Res. Natl. Inst. Stand. Technol. **103**, 259 (1998).

[28] L. Tian and P. Zoller, *Coupled Ion-Nanomechanical Systems*, Phys. Rev. Lett. **93**, 266403 (2004).

[29] W. K. Hensinger, D. W. Utami, H.-S. Goan, K. Schwab, C. Monroe, and G. J. Milburn, *Ion trap transducers for quantum electromechanical oscillators*, Phys. Rev. A **72**, 041405 (2005).

[30] D. Meiser and P. Meystre, *Coupled dynamics of atoms and radiation-pressure-driven interferometers*, Phys. Rev. A **73**, 033417 (2006).

[31] P. Treutlein, D. Hunger, S. Camerer, T. W. Hänsch, and J. Reichel, *Bose-Einstein Condensate Coupled to a Nanomechanical Resonator on an Atom Chip*, Phys. Rev. Lett. **99**, 140403 (2007).

[32] S. Singh, M. Bhattacharya, O. Dutta, and P. Meystre, *Coupling nanomechanical cantilevers to dipolar molecules*, Phys. Rev. Lett. **101**, 263603 (2008).

[33] C. Genes, D. Vitali, and P. Tombesi, *Emergence of atom-light-mirror entanglement inside an optical cavity*, Phys. Rev. A **77**, 050307 (2008).

[34] H. Ian, Z. R. Gong, Y. Liu, C. P. Sun, and F. Nori, *Cavity optomechanical coupling assisted by an atomic gas*, Phys. Rev. A **78**, 013824 (2008).

[35] A. A. Geraci and J. Kitching, *Ultracold mechanical resonators coupled to atoms in an optical lattice*, Phys. Rev. A **80**, 032317 (2009).

[36] A. B. Bhattacherjee, *Cavity quantum optomechanics of ultracold atoms in an optical lattice: Normal-mode splitting*, Phys. Rev. A **80**, 043607 (2009).

[37] K. Hammerer, M. Aspelmeyer, E. S. Polzik, and P. Zoller, *Establishing Einstein-Podolsky-Rosen Channels between Nanomechanics and Atomic Ensembles*, Phys. Rev. Lett. **102**, 020501 (2009).

[38] K. Hammerer, M. Wallquist, C. Genes, M. Ludwig, F. Marquardt, P. Treutlein, P. Zoller, J. Ye, and H. J. Kimble, *Strong coupling of a mechanical oscillator and a single atom*, Phys. Rev. Lett. **103**, 063005 (2009).

[39] K. Hammerer, K. Stannigel, C. Genes, P. Zoller, P. Treutlein, S. Camerer, D. Hunger, and T. W. Hänsch, *Optical Lattices with Micromechanical Mirrors*, Phys. Rev. A **82**, 021803.

[40] K. Zhang, W. Chen, M. Bhattacharya, and P. Meystre, *Hamiltonian chaos in a coupled BEC-optomechanical-cavity system*, Phys. Rev. A **81**, 013802 (2010).

[41] Y.-J. Wang, M. Eardley, S. Knappe, J. Moreland, L. Hollberg, and J. Kitching, *Magnetic Resonance in an Atomic Vapor Excited by a Mechanical Resonator*, Phys. Rev. Lett. **97**, 227602 (2006).

[42] W. Hänsel, P. Hommelhoff, T. W. Hänsch, and J. Reichel, *Bose-Einstein condensation on a microelectric chip*, Nature **413**, 498 (2001).

[43] H. Ott, J. Fortágh, G. Schlotterbeck, A. Grossmann, and C. Zimmermann, *Bose-Einstein Condensation in a Surface Microtrap*, Phys. Rev. Lett. **87**, 230401 (2001).

Bibliography

[44] Y. Colombe, T. Steinmetz, G. Dubois, F. Linke, D. Hunger, and J. Reichel, *Strong atom-field coupling for Bose-Einstein condensates in an optical cavity on a chip*, Nature **450**, 272 (2007).

[45] S. Hofferberth, I. Lesanovsky, B. Fischer, J. Verdu, and J. Schmiedmayer, *Radio-frequency dressed state potentials for neutral atoms*, Nature Physics **2**, 710 (2006).

[46] S. Hofferberth, I. Lesanovsky, T. Schumm, A. Imambekov, V. Gritsev, E. Demler, and J. Schmiedmayer, *Probing quantum and thermal noise in an interacting many-body system*, Nature Physics **4**, 489 (2008).

[47] P. Treutlein, P. Hommelhoff, T. Steinmetz, T. W. Hänsch, and J. Reichel, *Coherence in Microchip Traps*, Phys. Rev. Lett. **92**, 203005 (2004).

[48] Y. Lin, I. Teper, C. Chin, and V. Vuletic, *Impact of the Casimir-Polder Potential and Johnson Noise on Bose-Einstein Condensate Stability Near Surfaces*, Phys. Rev. Lett. **92**, 050404 (2004).

[49] W. Ketterle, D. S. Durfee, and D. M. Stamper-Kurn, *Making and probing and understanding Bose-Einstein condensates*, cond-mat/9904034 (1999).

[50] C. W. Gardiner, *Driving a quantum system with the output field from another driven quantum system*, Phys. Rev. Lett. **70**, 2269–2272 (1993).

[51] H. J. Carmichael, *Quantum trajectory theory for cascaded open systems*, Phys. Rev. Lett. **70**, 2273–2276 (1993).

[52] Q. P. Unterreithmeier, E. M. Weig, and J. P. Kotthaus, *Universal transduction scheme for nanomechanical systems based on dielectric forces*, Nature **458**, 1001 (2009).

[53] D. R. Koenig, E. M. Weig, and J. P. Kotthaus, *Ultrasonically driven nanomechanical single-electron shuttle*, Nature Nanotechnology **3**, 482 (2008).

[54] D. Hunger, *A Bose-Einstein condensate coupled to a micromechanical oscillator*, Dissertation, Ludwig-Maximilians-Universität, München, 2010.

[55] A. N. Cleland, *Foundation of Nanomechanics*, Springer, 2003.

[56] K. Ekinci and M. Roukes, *Nanoelectromechanical systems*, Rev. Sci. Instrum. **76**, 061101 (2005).

[57] M. J. Madou, *Fundamentals of Microfabrication: The Science of Miniaturization*, CRC Press and New York, 2nd edition (2002).

[58] J. D. Thompson, B. M. Zwickl, A. M. Jayich, F. Marquardt, S. M. Girvin, and J. G. E. Harris, *Strong dispersive coupling of a high-finesse cavity to a micromechanical membrane*, Nature **452**, 06715 (2008).

[59] S. S. Verbridge, J. M. Parpia, R. B. Reichenbach, L. M. Bellan, and H. G. Craighead, *High quality factor resonance at room temperature with nanostrings under high tensile stress*, Journal of Applied Physics **99**, 124304 (2006).

[60] W. Weaver, S. P. Timoshenko, D. H. Young, and eds., *Vibration Problems in Engineering*, John Wiley and Sons, 1990.

[61] G. Anetsberger, O. Arcizet, Q. P. Unterreithmeier, R. Riviere, A. Schliesser, E. M. Weig, J. P. Kotthaus, and T. J. Kippenberg, *Near-field cavity optomechanics with nanomechanical oscillators*, Nature Physics **5**, 909 (2009).

[62] O. Arcizet, P.-F. Cohadon, T. Briant, M. Pinard, and A. Heidmann, *Radiation-pressure cooling and optomechanical instability of a micromirror*, Nature **444**, 71 (2006).

[63] K. Yum, Z. Wang, A. P. Suryavanshi, and M.-F. Yu, *Experimental measurement and model analysis of damping effect in nanoscale mechanical beam resonators in air*, J. Appl. Phys. **96**, 3933 (2004).

[64] A. Cleland and M. Roukes, *Noise processes in nanomechanical resonators*, J. Appl. Phys. **92**, 2758 (2002).

[65] D. W. Carr, S. Evoy, L. Sekaric, J. M. Parpia, and H. G. Craighead, *Measurement of mechanical resonance and losses in nanometer scale silicon wires*, Appl. Phys. Lett. **75**, 920 (1999).

[66] R. Lifshitz and M. L. Roukes, *Thermoelastic damping in micro- and nanomechanical systems*, Phys. Rev. B **61**, 5600 (2000).

[67] D. J. Wilson, C. A. Regal, S. B. Papp, and H. J. Kimble, *Cavity Optomechanics with Stoichiometric SiN Films*, Phys. Rev. Lett. **103**, 207204 (2009).

[68] D. M. Photiadis and J. A. Judge, *Attachment losses of high Qoscillators*, Appl. Phys. Lett. **85**, 482 (2004).

[69] M. C. Cross and R. Lifshitz, *Elastic wave transmission at an abrupt junction in a thin plate with application to heat transport and vibrations in mesoscopic systems*, Phys. Rev. B **64**, 085324 (2001).

[70] I. Wilson-Rae, *Intrinsic dissipation in nanomechanical resonators due to phonon tunneling*, Physical Review B **77**, 245418 (2008).

[71] P. W. Anderson, B. I. Halperin, and C. M. Varma, *Anomalous Low-temperature Thermal Properties of Glasses and Spin Glasses*, The Philosophical Magazine **25**, 1 (1971).

[72] O. Arcizet, R. Riviere, A. Schliesser, G. Anetsberger, and T. J. Kippenberg, *Cryogenic properties of optomechanical silica microcavities*, Physical Review A **80**, 021803 (2009).

[73] D. R. Southworth, R. A. Barton, S. S. Verbridge, B. Ilic, A. D. Fefferman, H. G. Craighead, and J. M. Parpia, *Stress and Silicon Nitride: A Crack in the Universal Dissipation of Glasses*, Phys. Rev. Lett. **102**, 225503 (2009).

[74] J. Yang, T. Ono, and M. Esashi, *Surface effects and high quality factors in ultrathin single-crystal silicon cantilevers*, Appl. Phys. Lett. **77**, 3860 (2000).

[75] D. T. Gillespie, *The mathematics of Brownian motion and Johnson noise*, Am. J. Phys. **64**, 225 (1996).

[76] T. J. Kippenberg and K. J. Vahala, *Cavity Optomechanics: Back-Action at the Mesoscale*, Science **321**, 1172 (2008).

[77] T. Corbitt, C. Wipf, T. Bodiya, D. Ottaway, D. Sigg, N. Smith, S. Whitcomb, and N. Mavalvala, *Optical Dilution and Feedback Cooling of a Gram-Scale Oscillator to 6.9 mK*, Phys. Rev. Lett. **99**, 160801 (2007).

[78] R. Penrose, *On Gravity's Role in Quantum State Reduction*, General Relativity and Gravitation **28**, 581 (1996).

[79] W. Marshall, C. Simon, R. Penrose, and D. Bouwmeester, *Towards Quantum Superpositions of a Mirror*, Phys. Rev. Lett. **91**, 130401 (2003).

[80] M. D. LaHaye, J. Suh, P. M. Echternach, K. C. Schwab, and M. L. Roukes, *Nanomechanical measurements of a superconducting qubit*, Nature **459**, 960 (2009).

[81] D. Hunger, S. Camerer, T. W. Hänsch, D. König, J. P. Kotthaus, J. Reichel, and P. Treutlein, *Resonant Coupling of a Bose-Einstein Condensate to a Micromechanical Oscillator*, Phys. Rev. Lett. **104**, 143002 (2010).

[82] S. Camerer, M. Korppi, D. Hunger, A. Jöckel, T. Hänsch, and P. Treutlein, *Observation of backaction of ultracold atoms onto a mechanical oscillator*, to be submitted.

[83] D. A. Steck, *Rubidium 87 D Line Data*, http://steck.us/alkalidata/, version 1.6 (2003).

[84] D. Z. Anderson and J. G. J. Reichel, *Cold atom system with atom chip wall*, United States Patent 7126112 (2006).

[85] J. Reichel, *Microchip traps and BoseEinstein condensation*, Appl. Phys. B **75**, 469 (2002).

[86] J. Reichel, W. Hnsel, P. Hommelhoff, and T. Hänsch, *Applications of integrated magnetic microtraps*, Appl. Phys. B **72**, 81 (2001).

[87] R. Folman, P. Krüger, J. Schmiedmayer, J. Denschlag, and C. Henkel, *Microscopic Atom Optics: From Wires to an Atom Chip*, Adv. At. Mol. Opt. Phys. **48**, 263 (2002).

Bibliography

[88] J. Fortagh and C. Zimmermann, *Magnetic microtraps for ultracold atoms*, Rev. Mod. Phys. **79**, 235 (2007).

[89] D. M. Farkas, K. M. Hudek, E. A. Salim, S. R. Segal, M. B. Squires, and D. Z. Anderson, *A compact, transportable, microchip-based system for high repetition rate production of Bose–Einstein condensates*, Appl. Phys. Lett. **9**, 093102 (2010).

[90] R. Schmied, D. Leibfried, R. J. C. Spreeuw, and S. Whitlock, *Optimized magnetic lattices for ultracold atomic ensembles*, arXiv:1006.4532 (2010).

[91] T. Steinmetz, *Resonator-Quantenelektrodynamik auf einem Mikrofallenchip*, Dissertation, Ludwig-Maximilians-Universität, München, 2008.

[92] M. Greiner, O. Mandel, T. Esslinger, T. Hänsch, and I. Bloch, *Quantum Phase Transition from a Superfluid to a Mott Insulator in a Gas of Ultracold Atoms*, Nature **415**, 39 (2002).

[93] G. Jo, Y. Lee, J. Choi, C. Christensen, T. Kim, J. Thywissen, D. Pritchard, and W. Ketterle, *Itinerant ferromagnetism in a Fermi gas of ultracold atoms*, Science **325**, 1521 (2009).

[94] R. Grimm, M. Weidemüller, and Y. Ovchinnikov, *Optical dipole traps for neutral atoms*, Adv. At. Mol. Opt. Phys. **42**, 95 (2000).

[95] J. Jackson, *Classical electrodynamics*, Wiley, New York, 1962.

[96] W. Alt, D. Schrader, S. Kuhr, M. Müller, V. Gomer, and D. Meschede, *Single atoms in a standing-wave dipole trap*, Physical Review A **67**, 033403 (2003).

[97] S. L. Winoto, M. T. DePue, N. E. Bramall, and D. S. Weiss, *Laser cooling at high density in deep far-detuned optical lattices*, Phys. Rev. A **59**, R19 (1999).

[98] M. T. DePue, C. McCormick, S. L. Winoto, S. Oliver, and D. S. Weiss, *Unity Occupation of Sites in a 3D Optical Lattice*, Phys. Rev. Lett. **82**, 2262 (1999).

[99] A. J. Kerman, V. Vuletic, C. Chin, and S. Chu, *Beyond Optical Molasses: 3D Raman Sideband Cooling of Atomic Cesium to High Phase-Space Density*, Phys. Rev. Lett. **84**, 439 (2000).

[100] M. H. Anderson, W. Petrich, J. R. Ensher, and E. A. Cornell, *Reduction of light-assisted collisional loss rate from a low-pressure vapor-cell trap*, Phys. Rev. A **50**, R3597 (1994).

[101] H. Casimir and D. Polder, *The Influence of retardation on the London-van der Waals Forces*, Physical Review **73**, 362 (1948).

[102] M. Antezza, L. P. Pitaevskii, and S. Stringari, *Effect of the Casimir-Polder force on the collective oscillations of a trapped Bose-Einstein condensate*, Physical Review A **70**, 053610 (2004).

[103] Z.-C. Yan, A. Dalgarno, and J. F. Babb, *Long-range interactions of lithium atoms*, Physical Review A **55**, 2882 (1997).

[104] F. Zhou and L. Spruch, *van der Waals and retardation Casimir interactions of an electron or an atom with multilayered walls*, Physical Review A **52**, 297 (1995).

[105] S. Y. Buhmann, D.-G. Welsch, and T. Kampf, *Ground-state van der Waals forces in planar multilayer magnetodielectrics*, Physical Review A **72**, 032112 (2005).

[106] A. M. C. Reyes and C. Eberlein, *Casimir-Polder interaction between an atom and a dielectric slab*, Physical Review A **80**, 032901 (2009).

[107] R. D. Diehl and R. McGrath, *Current progress in understanding alkali metal adsorption on metal surfaces*, J. Phys.: Condens. Matter **9**, 951 (1997).

[108] F. Dalfovo, S. Giorgini, L. P. Pitaevskii, and S. Stringari, *Theory of Bose-Einstein condensation in trapped gases*, Reviews of Modern Physics **71**, 463 (1999).

[109] A. J. Leggett, *Bose-Einstein condensation in the alkali gases: Some fundamental concepts*, Reviews of Modern Physics **37**, 307 (2001).

[110] A. M. Mateo and V. Delgado, *Extension of the Thomas-Fermi approximation for trapped Bose-Einstein condensates with an arbitrary number of atoms*, Physical Review A **74**, 065602 (2006).

[111] S. Stringari, *Collective Excitations of a Trapped Bose-Condensed Gas*, Phys. Rev. Lett. **77**, 2360 (1996).

[112] T. Kimura, H. Saito, and M. Ueda, *A Variational Sum-Rule Approach to collective Excitations of a trapped Bose-Einstein Condensate*, Journal of the Physical Society of Japan **68**, 1477 (1999).

[113] P. Treutlein, *Coherent manipulation of ultracold atoms on atom chips*, Dissertation, Ludwig-Maximilians-Universität, München, 2008.

[114] J. T. Hoffrogge, *Mikrowellenpotentiale auf Atomchips*, Diplomarbeit (2007).

[115] Y. ju Lin, I. Teper, C. Chin, and V. Vuletic, *Impact of the Casimir-Polder Potential and Johnson Noise on Bose-Einstein Condensate Stability Near Surfaces*, Phys. Rev. Lett. **92**, 050404 (2004).

[116] J.-Y. Courtois, J.-M. Courty, and J. C. Mertz, *Internal dynamics of multilevel atoms near a vacuum-dielectric interface*, Phys. Rev. A **53**, 1862 (1996).

[117] F. J. Giessibl, *Advances in atomic force microscopy*, Reviews of Modern Physics **75**, 949 (2003).

[118] N. V. Morrow, S. K. Dutta, and G. Raithel, *Feedback Control of Atomic Motion in an Optical Lattice*, Phys. Rev. Lett. **88**, 093003 (2002).

Bibliography

[119] G. Raithel, W. D. Phillips, and S. L. Rolston, *Collapse and Revivals of Wave Packets in Optical Lattices*, Phys. Rev. Lett. **81**, 3615 (1998).

[120] M. Weidemüller, A. Görlitz, T. W. Hänsch, and A. Hemmerich, *Local and global properties of light-bound atomic lattices investigated by Bragg diffraction*, Phys. Rev. A **58**, 4647–4661 (1998).

[121] B. Zink and F. Hellman, *Specific heat and thermal conductivity of low-stress amorphous SiN membranes*, Solid State Communications **129**, 199204 (2004).

[122] T. Kwaaitaal, B. J. Luymes, and G. A. van der Pijll, *Noise limitations of Michelson laser interferometers*, J. Phys. D:Appl. Phys. **13**, 1005 (1980).

[123] L. Ricci, M. Weidemüller, T. Esslinger, A. Hemmerich, C. Zimmermann, V. Vuletic, W. König, and T. Hänsch, *A compact grating-stabilized diode laser system for atomic physics*, Optics Communications **117**, 541 (1995).

[124] W. Demtröder, *Experimentalphysik 2*, Springer, 5. Auflage (2008).

[125] M. Mader, *Characterisation of Silicon Nitride Membranes for Optomechanical Experiments*, Bachelor thesis, LMU München (2010).

[126] I. Wilson-Rae, R. Barton, S. Verbridge, D. Southworth, B. Ilic, H. Craighead, and J. Parpia, *High-Q Nanomechanics via Destructive Interference of Elastic Waves*, arXiv:1010.217v1 (2010).

[127] H. Steck, M. Naraschewski, and H. Wallis, *Output of a Pulsed Atom Laser*, Phys. Rev. Lett. **80**, 1 (1998).

[128] X. Zhua, P. Grütter, V. Metlushko, Y. Hao, F. J. Castano, C. A. Ross, B. Ilic, and H. Smith, *Construction of hysteresis loops of single domain elements and coupled permalloy ring arrays by magnetic force microscopy*, Journal of Applied Physics **93**, 8540 (2003).

[129] D. Jiles, *Recent advances and future directions in magnetic materials*, Acta Materialia **51**, 59075939 (2003).

[130] L. Kong and S. Y. Chou, *Effects of bar length on switching field of nanoscale nickel and cobalt bars fabricated using lithography*, J. Appl. Phys **80**, 5205 (1996).

[131] R. Engel-Herbert and T. Hesjedal, *Calculation of the magnetic stray field of a uniaxial magnetic domain*, J. Appl. Phys. **97**, 074504 (2005).

[132] S. Camerer, *Fabrikation und Charakterisierung eines nanomechanischen Systems zur Ankopplung an ein Bose-Einstein Kondensat*, Diplomarbeit, Ludwig-Maximilians-Universität / Technical University of Munich, München, 2006.

[133] W. Yao, T. A. Knuuttila, K. K. Nummila, J. E. Martikainen, A. S. Oja, and O. V. Lounasmaa, *A Versatile Nuclear Demagnetization Cryostat for Ultralow Temperature Research*, Journal of Low Temperature Physics **120**, 121 (2000).

[134] I. Favero, C. Metzger, S. Camerer, D. Koenig, H. Lorenz, J. P. Kotthaus, and K. Karrai, *Optical cooling of a micromirror of wavelength size*, Applied Physics Letters **90**, 104101 (2007).

[135] X. Li and P. W. Bohn, *Metal-assisted chemical etching in HF/H_2O_2 produces porous silicon*, Applied Physics Letters **77**, 2572 (2000).

[136] N. Megouda, T. Hadjersi, O. Elkechai, R. Douani, and L. Guerbous, *Bi-assisted chemical etching of silicon in $HF/CoNO_{3,2}$ solution*, Journal of Luminescence **129**, 221 (2009).

[137] M. L. Chouroua, K. Fukamia, T. Sakkaa, S. Virtanenc, and Y. H. Ogataa, *Metal-assisted etching of p-type silicon under anodic polarization in HF solution with and without H_2O_2*, Electrochimica Acta **55**, 903 (2010).

[138] S.-W. Chang, V. P. Chuang, S. T. Boles, C. A. Ross, and C. V. Thompson, *Densely Packed Arrays of Ultra-High-Aspect-Ratio Silicon Nanowires Fabricated using Block-Copolymer Lithography and Metal-Assisted Etching*, Advanced Functional Materials **19**, 2495 (2009).

[139] M. Poggio and C. Degen, *Force-detected nuclear magnetic resonance: recent advances and future challenges*, Nanotechnology **21**, 342001 (2010).

[140] A. C. Bleszynski-Jayich, W. E. Shanks, B. R. Ilic, and J. G. E. Harris, *High sensitivity cantilevers for measuring persistent currents in normal metal rings*, J. Vac. Sci. Technol. B **26**, 1412 (2008).

[141] A. K. Hüttel, G. A. Steele, B. Witkamp, M. Poot, L. P. Kouwenhoven, and H. S. J. van der Zant, *Carbon Nanotubes as Ultrahigh Quality Factor Mechanical Resonators*, Nano Letters **9**, 2547 (2009).

[142] P. Poncharal, Z. L. Wang, D. Ugarte, and W. A. de Heer, *Electrostatic Deflections and Electromechanical Resonances of Carbon Nanotubes*, Science **283**, 1513 (1999).

[143] R. Fermani, S. Scheel, and P. L. Knight, *Trapping cold atoms near carbon nanotubes: Thermal spin flips and Casimir-Polder potential*, Phys. Rev. A **75**, 062905 (2007).

[144] M. Mücke, E. Figueroa, J. Bochmann, C. Hahn, K. Murr, S. Ritter, C. Villas-Boas, and G. Rempe, *Electromagnetically induced transparency with single atoms in a cavity*, Nature **465**, 755 (2010).

[145] P. Rabl, P. Cappellaro, M. V. G. Dutt, L. Jiang, J. R. Maze, and M. D. Lukin, *Strong magnetic coupling between an electronic spin qubit and a mechanical resonator*, Phys. Rev. B **79**, 041302 (2009).

Dank

Hilfreich und gut ist es, angelegentlich aus dem Strom der Zeit ans Ufer zu treten und dort innehaltend der verflossenen Zeit nachzublicken. Es zeigt sich, daß auf den vorangehenden Seiten vor allem die Früchte der Arbeit, nicht aber der Nährboden beschrieben ist.

Für die Aufnahme in seine Arbeitsgruppe danke Prof. Hänsch. Das diskussionsfreudige und vertrauensvolle Klima ist eine ideale Vorraussetzung für wissenschaftlich geordneten Wildwuchs. Besonders möchte ich mich für die Möglichkeit bedanken, die ICAP 2006, die Les Houches Summer School in Singapore 2009 und die ICAP 2010 zu besuchen.

Vor allen danke ich meinem Betreuer Prof. Philipp Treutlein, der meine Arbeit durch zahllose Anregungen wie kein anderer bereichert hat. Neben der wissenschaftlichen Betreuung genoß ich vor allem das freundschaftliche Verhältnis, das sich in gemeinsamer Freizeitgestaltung, vom Chorgesang bis hin zu gemeinsamem Rennradfahren manifestierte.

Herzlich danke ich David Hunger: Er teilte mit mir während dreier Jahre die tägliche Arbeit im Labor und die Gummibärchen. Viel verdanke ich seiner stoischen Ruhe und beinahe protestantischen Arbeitsethik, die wohltuend vom epikureischen Zug des Käse-, Schokolade-, Rennrad-, und Weingenießers durchsetzt war. Es hat mir gutgetan die letzten Jahre in seiner Nähe zu verbringen.

Ebenso herzlich danke ich Maria Korppi, die mit mir etwa ein Jahr lang im Labor werkte und meine Gummibärchen mit mir teilte. Neben gemeinsamen kulinarischen Ausflügen zu amerikanischen Restaurants unternahmen wir Rennradtouren in die Berge und höllische Ruderpartien in der Sommerhitze.

Matthias Mader baute mit großem technischem und ästhetischem Feingefühl ein Experiment auf, und bereicherte unser Leben im Labor: herzlichen Dank dafür! Trotz konfessioneller Diskrepanzen war die Diskussion über den Papst von großer Fairness und gegenseitigem Verständnis geprägt: wir haben den Maßkrug nur ganz leicht auf'gesetzt.

Herzlich danke ich Andreas Jöckel, der in unsere Gruppe kam, als ich schon fast begonnen hatte, meine Arbeit zu Papier zu bringen. Er bereicherte unsere Diskussionen durch scharfsinnige Einfälle, und den Laboralltag durch seinen trockenen Humor.

Ganz besonders möchte ich Maria und Andreas für die sorgfältige Fortführung der Experimente nach meinem Ausscheiden aus dem Labor, und die hingebungsvolle Auswertung der Meßdaten danken.

Ich danke meinen Kollegen vom Mikrowellenexperiment Pascal Böhi, Max Riedel, Johannes Hoffrogge, Jad Halimeh und Roman Schmied für zahllose Diskussionen und gemeinsame Unternehmungen ausserhalb des Labores.

Ich danke allen Mitgliedern der Arbeitsgruppen Hänsch und Weinfurter, insbesondere Hannes Brachmann, Luis Costa, Carolin Hahn, Florian Henkel, Michael Krug und Daniel Schlenk für die angenehme Atmosphäre in der Schellingstraße.

Ich danke Prof. Kotthaus für die großzügige Erlaubnis, den Reinraum an seinem Lehrstuhl zu nutzen, und schließlich für die Erstellung des Zweitgutachtens meiner Arbeit. Auch Daniel König, Quirin Unterreithmeier, Alexander Paul, Stefan Schöffberger, Constanze Höhberger Metzger, Ivan Favero, Sebastian Stapfner, und Bert Lorenz danke ich für viele fruchtbare Diskussionen.

Ich danke Prof. Jakob Reichel für die Einladungen zu den pariser Paris-Munich Atomchip Konferenzen, die jeweils mit einem wunderbaren Parisaufenthalt verbunden waren.

Toni Scheich danke ich für viele gute Ratschläge und Hilfe bei meinen Lötprojekten. Ich danke Herrn Aust und Herrn Grosshauser von der LMU Werkstatt, sowie Wolfgang Simon und Charly Linner von der MPQ Werkstatt für schnelle, unkomplizierte Hilfe.

Schließlich möchte ich Gabriele Gschwendtner für ihr lautloses und effizientes Wirken im Hintergrund, die reibungslose Organisation von Konferenzen, Vertragsverlängerungen, etc. danken! Nicole Schmidt danke ich für Hilfe bei der Nutzung des Chemielabores und die Bereithaltung von stets aktualisierten Softwarepaketen.

Ganz besonders aber möchte ich meinen Eltern für ihre immerwährende Unterstützung danken.

Die VDM Verlagsservicegesellschaft sucht für wissenschaftliche Verlage abgeschlossene und herausragende

Dissertationen, Habilitationen, Diplomarbeiten, Master Theses, Magisterarbeiten usw.

für die kostenlose Publikation als Fachbuch.

Sie verfügen über eine Arbeit, die hohen inhaltlichen und formalen Ansprüchen genügt, und haben Interesse an einer honorarvergüteten Publikation?

Dann senden Sie bitte erste Informationen über sich und Ihre Arbeit per Email an *info@vdm-vsg.de*.

Sie erhalten kurzfristig unser Feedback!

VDM Verlagsservicegesellschaft mbH
Dudweiler Landstr. 99
D - 66123 Saarbrücken

Telefon +49 681 3720 174
Fax +49 681 3720 1749

www.vdm-vsg.de

Die VDM Verlagsservicegesellschaft mbH vertritt

Printed by Books on Demand GmbH, Norderstedt / Germany